Stefan Wippermann

Understanding quasi-1D atomic-scale nanowires from ab initio theory

Stefan Wippermann

Understanding quasi-1D atomic-scale nanowires from ab initio theory

Electron transport, optical and thermodynamical properties

Südwestdeutscher Verlag für Hochschulschriften

Imprint
Any brand names and product names mentioned in this book are subject to trademark, brand or patent protection and are trademarks or registered trademarks of their respective holders. The use of brand names, product names, common names, trade names, product descriptions etc. even without a particular marking in this work is in no way to be construed to mean that such names may be regarded as unrestricted in respect of trademark and brand protection legislation and could thus be used by anyone.

Publisher:
Südwestdeutscher Verlag für Hochschulschriften
is a trademark of
Dodo Books Indian Ocean Ltd., member of the OmniScriptum S.R.L Publishing group
str. A.Russo 15, of. 61, Chisinau-2068, Republic of Moldova Europe
Printed at: see last page
ISBN: 978-3-8381-1877-2

Zugl. / Approved by: Paderborn, Universität, Dissertation, 2010

Copyright © Stefan Wippermann
Copyright © 2010 Dodo Books Indian Ocean Ltd., member of the OmniScriptum S.R.L Publishing group

Stefan Wippermann, *Understanding substrate-supported atomic-scale nanowires from ab initio theory*.
PhD Thesis (in English), Department of Physics, Faculty of Science, University of Paderborn, Germany (2010). 184 Pages, 69 Figures, 8 Tables.

Abstract

One-dimensional (1D) electronic systems are currently intensively investigated for both fundamental and technological reasons. With respect to future nanoelectronic device concepts the present work aims at generating a solid knowledge and detailed understanding about the physical foundations of such perspective future devices. Highly anisotropic surface superstructures have attracted considerable attention in this context. An acutely studied model system of this kind is the ordered atomic-scale array of self-assembled In nanowires that forms the Si(111)-(4×1)In phase at room temperature (RT) [25]. More than 10 years ago it was discovered that this nanowire array undergoes a reversible phase transition from (4×1)→(8×2) translational symmetry at T_C = 120 K [30]. However, while being discussed intensively in the scientific literature, both the (4×1)→(8×2) phase transition's driving mechanism and low temperature (LT) ground-state with its associated properties remain strongly controversial.

In the present work the In/Si(111)-(4×1)/(8×2) nanowire array is investigated by means of state-of-the-art *ab initio* computer simulations. It is demonstrated that the longstanding problem of determining the internal structure and exact electronic properties of the nanowire array's LT ground-state cannot be resolved by the surface energetics alone. It turns out that the density functional theory (DFT) total-energy results for the In/Si(111)-(4×1)/(8×2) surface are extremely sensitive with respect to the details of the electron-electron interaction treatment. Electronic structure and transport calculations performed for the trimer and hexagon models of the LT ground-state indicate hexagon formation, as first suggested by González et al. [35]. These results demonstrate the distinct influence of small changes of the nanowire geometry on its conductance (cf. Publ. [14]). Given the ambiguities of the total-energy calculations in determining the internal structure of the (8×2) ground-state, the comparison of optical fingerprints calculated for structural candidates with measured data is expected to be helpful. Calculations of the anisotropic optical response in the visible and mid-infrared regime including intraband transitions have been performed for the In/Si(111)-(4×1)/(8×2) nanowire array for the first time. It is demonstrated that states close to the Fermi energy lead to distinct and unique optical fingerprints in the mid-infrared regime for each of the examined structural models. Only the spectra of the (8×2) hexagon model agree closely with recent measurements. These results are suitable to effectively conclude the search that has been ongoing for more than 10 years (cf. Publ. [3,8,9]). To address the driving mechanism of the phase transition the In/Si(111)-(4×1)/(8×2) surface's thermal properties have been explored by large-scale frozen phonon and molecular dynamics (MD) simulations. The results indicate that the soft shear mode mechanism, as proposed by González et al. [35], is at least partially correct. Two further soft phonon modes in conjunction with the shear mode facilitate the phase transition. By comparing the present results to the Raman spectroscopy measurements by Fleischer et al. [46] the existence of these soft modes could be confirmed for the first time.

In transport physics calculations and measurements are usually carried out at low bias, low temperature since the main focus is centered on fundamental principles. However, with respect to device applications the high temperature properties are most important. Based on the (8×2) hexagon model for the LT ground-state a combined frozen phonon and MD approach is presented to derive the temperature dependent transport properties including the phase transition. However, the classical energy distribution employed by MD calculations effectively prohibits a sufficiently accurate treatment of the highly subtle energetics of the In/Si(111)-(4×1)/(8×2) nanowire array. Instead a quantum Monte Carlo approach is proposed that incorporates both the correct potential energy landscape and energy distribution. As doping represents the basic building block of today's microelectronics *first principles* calculations of the Landauer conductance of doped In nanowires have been performed for various adatom species. Distinct conductance modifications are predicted that are related to potential well scattering, nanowire deformation or a combination of both effects. (cf. Publ. [11]).

Keywords

density functional theory, DFT, low dimensional systems, nanowires, defects, surface states, electronic structure, electron transport, optical properties, lattice dynamics

Stefan Wippermann, *Understanding substrate-supported atomic-scale nanowires from ab initio theory*.
Dissertation (in englischer Sprache), Dep. Physik, Fakultät für Naturwissenschaften, Universität Paderborn, Deutschland (2010). 184 Seiten, 69 Abbildungen, 8 Tabellen.

Kurzfassung

Systeme mit ein-dimensionaler (1D) Elektronenstruktur werden zur Zeit intensiv erforscht, sowohl aufgrund fundamentaler als auch technologischer Interessen. Die vorliegende Arbeit zielt auf die Erlangung soliden Wissens und detaillierten Verständnisses der physikalischen Grundlagen zukünftiger nanoelektronischer Bauelementkonzepte. Stark anisotrope Oberflächenüberstrukturen erhalten derzeit große Aufmerksamkeit in diesem Kontext. Bei der atomar-skaligen Anordnung selbstorganisierter In Nanodrähte, welche die Si(111)-(4×1)In Phase bei Raumtemperatur (RT) bildet, handelt es sich um ein besonders intensiv erforschtes Modellsystem dieser Art [1]. Bereits vor 10 Jahren wurde erstmals ein reversibler Phasenübergang in dieser Nanodraht-Anordnung von (4×1)→(8×2) translationaler Symmetrie bei $T_C = 120K$ beobachtet [2]. Trotz fortwährender intensiver Diskussion in der wissenschaftlichen Literatur, verbleiben sowohl der Mechanismus des Phasenübergangs als auch der Niedrigtemperatur (NT) Grundzustand samt seiner Eigenschaften stark umstritten.

In der vorliegenden Arbeit wird das In/Si(111)-(4×1)/(8×2) Nanodraht-System mit Hilfe akkurater *ab initio* Computersimulationen untersucht. Es wird gezeigt, dass das lang bestehende Problem der Bestimmung der internen Struktur und elektronischen Eigenschaften des NT Grundzustands allein mit Hilfe der Oberflächenenergien nicht gelöst werden kann. Die mittels Dichtefunktional-Theorie (DFT) erhaltenen Gesamtenergien der In/Si(111)-(4×1)/(8×2) Oberfläche hängen stark von den Details der Behandlung der Elektron-Elektron Wechselwirkung ab. Berechnungen der Elektronenstruktur und -transport Eigenschaften für die Trimer und Hexagon Modelle des NT Grundzustands deuten auf Hexagonbildung hin, wie erstmals durch González *et al.* vorgeschlagen [3]. Diese Ergebnisse demonstrieren den ausgeprägten Einfluss geringfügiger Geometrieänderungen der Nanodrähte auf ihren Leitwert (vgl. Publ. [1]). Hinsichtlich der Mehrdeutigkeit der Gesamtenergie Rechnungen in der Bestimmung der Grundzustands-Struktur, sind weiterführende Ergebnisse durch einen Vergleich der optischen Fingerabdrücke struktureller Kandidaten mit dem Experiment zu erwarten. Das anisotrope optische Antwortverhalten der In/Si(111)-(4×1)/(8×2) Nanodraht-Anordnung wurde erstmals im sichtbaren und mittleren Infrarot Bereich unter Berücksichtigung von Intrabandübergängen berechnet. Es wird gezeigt, dass Zustände nahe der Fermi-Kante für jedes der untersuchten Strukturmodelle ausgeprägte und einzigartige optische Fingerabdrücke im mittleren Infrarot-Bereich hervorrufen. Ausschließlich die Spektren des (8×2) Hexagon Modells stimmen mit aktuellen Messungen überein. Diese Ergebnisse beschließen die mehr als 10 Jahre währende Suche nach dem NT Grundzustand überzeugend (vgl. Publ. [3,8,9]).

Zur Untersuchung des den Phasenübergang antreibenden Mechanismus' wurden die thermischen Eigenschaften der In/Si(111)-(4×1)/(8×2) Oberfläche durch frozen phonon und Molekulardynamik (MD) Simulationen untersucht. Die Ergebnisse deuten darauf hin, dass der von González et al. vorgeschlagene Mechanismus mittels einer weichen Scherungsmode [35] zumindest teilweise korrekt ist. Zusammen mit der Scherungsmode unterstützen zwei weitere weiche Moden den Phasenübergang. Vergleichend mit den Raman Spektroskopie Daten von Fleischer et al. [46] konnte die Existenz dieser weichen Moden erstmals bestätigt werden.

Transport Physik wird für gewöhnlich bei niedrigen Spannungen und niedrigen Temperaturen betrieben, da der Schwerpunkt auf den grundlegenden Prinzipien liegt. Hinsichtlich der Anwendung in Bauelementen sind jedoch auch die Hochtemperatur Eigenschaften von großer Wichtigkeit. Basierend auf dem (8×2) Hexagon Modell des NT Grundzustands wird ein kombinierter frozen phonon und MD Ansatz zur Berechnung der temperaturabhängigen Transporteigenschaften inklusive des Phasenübergangs vorgestellt. Es stellt sich allerdings heraus, dass die von der MD verwendete klassische Energieverteilung eine hinreichend genaue Beschreibung der subtilen Energetik der In/Si(111)-(4×1)/(8×2) Oberfläche verhindert. Statt dessen wird ein Quantum Monte Carlo Algorithmus vorgestellt, welcher sowohl die potentielle Energieoberfläche als auch die Energieverteilung korrekt beschreibt. Da das Dotieren den Grundbaustein der modernen Mikroelektronik darstellt, wurden außerdem Berechnungen der Landauer Leitfähigkeit für mit verschiedenen Fremdatomen dotierte In Nanodrähte durchgeführt. Es werden ausgeprägte Modifikationen der Leitfähigkeit vorhergesagt, welche durch Potentialtopf Streuung, Deformation der Nanodrähte oder eine Kombination beider Effekte erklärt werden können (vgl. [11]).

Schlagwörter

Dichtefunktional-Theorie, DFT, Niederdimensionale Systeme, Nanodrähte, Defekte, Oberflächenzustände, elektronische Struktur, Elektronen Transport, optische Eigenschaften, Gitterdynamik

Contents

Abstract I
List of tables VIII
List of figures IX

1 Introduction   **1**
    1.1 Motivation 1
    1.2 Previous Research 5
       1.2.1 Structure and phase transition 5
       1.2.2 Transport properties and doping 12
    1.3 Project outline 17

2 Methodology   **19**
    2.1 Density Functional Theory (DFT) 19
       2.1.1 The many body problem 19
       2.1.2 The Hohenberg-Kohn theorems 21
       2.1.3 The Kohn-Sham equations 23
       2.1.4 Exchange-correlation (XC) functionals 25
           2.1.4a Local density approximation (LDA) 26
           2.1.4b Generalized gradient approximation (GGA) 28
       2.1.5 Periodic boundary conditions 28
           2.1.5a Plane-wave basis set expansion 29
           2.1.5b Supercell method 31
           2.1.5c k-space integration 32
       2.1.6 The pseudopotential approach 34
           2.1.6a Frozen-core approximation 34
           2.1.6b Pseudopotential concept 35
           2.1.6c Norm-conserving pseudopotentials (NC-PP) 36
           2.1.6d Ultrasoft pseudopotentials (US-PP) 36

		2.1.6e	Projector-augmented wave pseudopotentials (PAW) . .	38
	2.1.7		Calculation of the forces: Hellmann-Feynman theorem	38
2.2	Transport theory			40
	2.2.1		Preliminary concepts	40
		2.2.1a	Subbands or transverse modes	40
		2.2.1b	Degenerate and non-degenerate conductors	43
		2.2.1c	Characteristic length scales	44
		2.2.1d	Conduction as dynamics of Fermi-energy electrons . .	45
	2.2.2		Landauer conductance formalism	46
		2.2.2a	Calculation of the contact resistance	47
		2.2.2b	The Landauer formula	48
		2.2.2c	Linear response for non-zero temperatures	50
	2.2.3		Landauer conductance from Green's functions	52
		2.2.3a	Green's function formalism	53
		2.2.3b	Fisher-Lee relation	56
		2.2.3c	Tight-binding approach	57
	2.2.4		Bridging the gap to DFT	59
		2.2.4a	Transmission within the supercell approach	60
		2.2.4b	Real-space basis set: Wannier functions	62
		2.2.4c	Localization procedure	63
		2.2.4d	Obtaining the real-space Hamiltonian	66
2.3	Optical and spectroscopic properties			68
	2.3.1		Reflectance anisotropy spectroscopy (RAS)	68
	2.3.2		Obtaining the dielectric tensor from DFT	69
		2.3.2a	Independent particle approximation (IPA)	70
		2.3.2b	Intraband contributions in metallic case	72
	2.3.3		Many-body correction Green's function schemes	76
		2.3.3a	One- and two-particle Green's functions	76
		2.3.3b	Electronic self-energy and Hedin's equations	77
		2.3.3c	Independent quasi-particle approximation (IQA)	80
		2.3.3d	Electron-hole attraction and Bethe-Salpeter Eq. (BSE) .	83
		2.3.3e	Implications on the bandstructure and dielectric tensor	86
2.4	Program packages (VASP, PWScf/WanT, DP)			90

3 **Transport properties of the clean Si(111)-(4×1)/(8×2)-In surface**   **93**
   3.1 Direct approaches to structure and why they fail 93
   3.2 Electronic properties 96

3.3	Transport properties		99
	3.3.1	Computational details	99
	3.3.2	Conductance spectra and discussion	103

4 Transport properties of the doped Si(111)-(4×1)/(8×2)-In surface — 105

- 4.1 Adsorption of impurity atoms on In/Si(111)-(4×1) ... 105
  - 4.1.1 Potential energy surfaces (PES) ... 105
  - 4.1.2 Structural properties ... 108
- 4.2 Transport properties ... 110
  - 4.2.1 Computational details ... 110
  - 4.2.2 Conductance spectra ... 112
  - 4.2.3 Adatom-localized phonon modes ... 113
- 4.3 Conductance quenching mechanisms ... 115
  - 4.3.1 Local density of states (DOS) ... 115
  - 4.3.2 Potential-well scattering ... 116
  - 4.3.3 Structural effects ... 118
  - 4.3.4 Discussion ... 119

5 Optical properties — 121

- 5.1 Optical anisotropy in the visible spectral range ... 121
  - 5.1.1 Computational details ... 122
  - 5.1.2 Results ... 123
- 5.2 Structure determination by mid-infrared response ... 126
  - 5.2.1 Computational details ... 126
  - 5.2.2 Mid-infrared optical anisotropy ... 127
  - 5.2.3 Transitions responsible for the observed anisotropy ... 129
- 5.3 Discussion ... 131

6 Thermal properties — 133

- 6.1 Phonon spectra in theory & experiment ... 133
  - 6.1.1 In/Si(111)-(4×1) surface ... 133
  - 6.1.2 In/Si(111)-(8×2) hexagon structure ... 138
- 6.2 Temperature-dependent transport properties ... 141
  - 6.2.1 *Frozen-phonon* (FP) approach ... 141
  - 6.2.2 *Molecular dynamics* (MD) approach ... 142
  - 6.2.3 A simple test system: zigzag Au wires ... 143
  - 6.2.4 Quasiparticle corrections and eigenstate symmetries ... 145
  - 6.2.5 Results of the combined FP and MD approaches ... 148

	6.3	Discussion	149
7	**Summary and conclusions**	**155**
	7.1	Results of the present work	155
	7.2	Outlook	161
		7.2.1 Quantum mechanical treatment of the phase transition	161
		7.2.2 Entropy contributions	162
		7.2.3 Doping vs. optical pumping	163
8	**Addendum**	**165**
	8.1	Entropy explains the phase transition	165
References		**171**
Publications		**187**
Acknowledgements		**189**

List of Tables

1.1 Atomic units used in the present work 21

1.1 Formation energies (in meV per (4×1) unit cell) of In/Si(111) surface reconstructions relative to the ideal (4×1) chain in dependence on the treatment of the electron exchange-correlation and the explicit inclusion of the In $4d$ states (cf. Publ. [14]). 95

1.1 Standard deviation σ of the In-In bond length distribution and average shift $\bar{\Delta}$ of the In chain atoms for ideal and defect-modified nanowires. . 109

2.2 Average quantum conductance G and DOS in the energy interval ±0.05 eV around the Fermi energy. 112

3.3 Average quantum conductance G in the energy interval ±0.05 eV around the Fermi energy. G' refers to the respective adatom structure without the adatom. 119

1.1 Calculated frequencies of A' and A'' surface vibrational Γ-point modes of the In/Si(111)-(4×1) surface 134

2.2 Average conductances and standard deviations in units of $2e^2/h$ in linear response for infinite zigzag Au chains at finite temperatures [197]. . 145

2.3 Average conductances in linear response and average displacements of the In atoms from their equilibrium positions within the Si(111)-(8×4)In hexagon model. 149

1.1 Calculated Γ-point frequencies of the (4×1)/(8×2) phases in comparison with experimental data. 166

List of Figures

1.1	Development of the speed increase per unit of cost for computing devices since 1900 [1, 2].	1
1.2	Cross section of a 45 nm microprocessor	2
1.3	Memristive crossbar circuit	4
2.4	Structural model of the In/Si(111)-(4×1) room temperature phase	6
2.5	STM images of the Si(111)-(4×1)/(8×2)-In surface at 121K	7
2.6	Structural models for the low and high temperature phases of In/Si(111)-(4×1)/(8×2)	8
2.7	Exp. vs. theoretical RA-spectra for the (4×1)/(8×2) RT/LT phase	11
2.8	Temperature dependent conductivity of the In/Si(111) surface	12
2.9	Temperature dependent conductivity for clean and 0.1ML In-decorated In nanowires	13
2.10	STM image of the (8×2) LT phase after deposition of 0.1ML In	15
2.11	Surface state dispersion and band bending upon Na deposition	15
2.12	STM image and pair correlation function of the Co adsorbed In/Si(111)-(4×1) surface	16
1.1	Supercell methods for symmetric and asymmetric slabs	32
1.2	Schematic illustration of all-electron and pseudo wavefunctions and potentials.	36
1.3	DFT calculational scheme	39
2.4	Fermi circle and net current flow at T = 0 K	45
2.5	Conductor setup and dispersion relation for the derivation of the Landauer formula	50
2.6	Energy distribution of the incident electrons at non-zero temperature.	51
2.7	Conductor setup in tight binding Greens function representation	58

2.8	Schematic representation of the left lead – conductor – right lead (LCR) system within the principal layer approach	61
3.9	Schematic representation of the Drude free electron gas dielectric function ($\omega_p^2 = 10^4 \text{s}^{-1}, \tau = 1\text{s}$).	74
3.10	Diagrammatic representation of the Bethe-Salpeter equation	83
3.11	Primitive unit cell of hexagonal LiNbO$_3$ in ferroelectric phase.	87
3.12	Band structures and dielectric functions of LiNbO$_3$ at GGA, GWA and BSE levels of theory, respectively	88
3.13	Notation of high symmetry Brillouin zone points.	88
1.1	Schematic top views of the different In/Si(111)-(4×1)/(8×2) structural models	94
2.2	Projected Si-bulk band structure	97
2.3	Total effective single particle potential and band structure of In/Si(111)-(4×1)	98
2.4	Band structures of the (4×2) and (8×2) trimer and hexagon models . .	99
3.5	Employed model system for the (4×1) ideal surface and orbital symmetries of Wannier trial centers	100
3.6	Comparison of band structures in plane wave and Wannier function basis sets	102
3.7	Quantum conductance spectra in dependence on the slab thickness . .	103
3.8	Quantum conductance spectra for electron transport along the chain direction calculated for In/Si(111) model structures (cf. Publ. [14]). . . .	104
1.1	Potential energy surfaces (PES) calculated for indium and lead adatoms on the In/Si(111)-(4×1) surface	106
1.2	Potential energy surfaces (PES) calculated for hydrogen and oxygen adatoms on the In/Si(111)-(4×1) surface	107
1.3	Schematic top and side views of In and Pb adatoms adsorbed on the In/Si(111)-(4×1) surface.	108
1.4	Schematic top and side views of H and O adatoms adsorbed on the In/Si(111)-(4×1) surface.	109
2.5	Partitioning of the *left lead – conductor – right lead* (LCR) system into a (8×8) conductor cell and separate (8×2) cells for the semi-infinite leads	110
2.6	Quantum conductance spectra and density of states for electron transport through ideal and adatom-modified In/Si(111) structures	112
2.7	Adatom-localized phonon modes and their impact on the quantum conductance for room temperature occupation	114

LIST OF FIGURES

3.8 Adsorption induced changes of the local density of states at E_F 116

3.9 Computational setup and amplitude distribution snapshots for numerically solving the 2-dim Schrödinger eq. for gaussian wave-packets of Fermi-energy electrons traveling towards the Pb-induced potential well 117

3.10 Averaged effective potential along the wire direction calculated for ideal and adatom-modified In/Si(111) structures 117

3.11 Quantum conductance spectra for electron transport along the wire direction calculated for ideal and adatom-modified In/Si(111) structures that do, however, not contain the adatom itself (cf. Publ. [11]). 118

3.12 Isodensity surface of the local DOS at E_F, illustrating the local DOS at the ideal In nanowires and its modification upon Pb adsorption. 120

1.1 RAS spectra calculated for the (4×1) ideal and (8×2) trimer and hexagon models of the In/Si(111) nanowire array, respectively (cf. Publ. [9]). . . 123

1.2 Squared wavefunctions of surface electronic states that strongly contribute to the optical anisotropy of the In/Si(111)-(4×1) nanowire array 124

1.3 Squared wavefunctions of surface electronic states that strongly contribute to the optical anisotropy of the hexagon model for the In/Si(111)-(8×2) nanowire array 125

2.4 Calculated and measured RA-spectra of Si(111)-(4×1)In at RT (300K) and Si(111)-(8×2)In at LT (70K) 128

2.5 Band structure and pronounced anisotropic optical transitions of the hexagon model for the In/Si(111)-(8×2) surface 130

2.6 Squared wavefunctions of surface electronic states at the X and M high-symmetry points that contribute strongly to the optical anisotropy of the In/Si(111)-(8×2) hexagon structure in the mid-IR regime 131

2.7 3-dim band structure and pronounced anisotropic optical transitions over the entire surface Brillouin zone of the In/Si(111)-(8×2) hexagon model 132

1.1 Calculated and measured phonon modes and their respective degrees of localization for the In/Si(111)-(4×1) surface 135

1.2 Eigenvectors of the soft shear and rotary modes, respectively 136

1.3 Linear decomposition of the (4×2) hexagon and trimer models into phonon eigenvectors. 137

1.4 Calculated and measured phonon modes and their respective degrees of localization for the In/Si(111)-(8×2) hexagon surface 138

1.5 Symmetric and antisymmetric soft shear modes of the (8×2) surface . . 139

1.6　Linear decomposition of the (4×1) ideal reconstruction into the phonon eigenvectors of the (8×2) hexagon structure. 140
2.7　Geometric and electronic structures of monoatomic Au zigzag chains . 143
2.8　Phonon dispersion relation of the zigzag Au chain 144
2.9　Original DFT-LDA and eigenstate symmetry sorted band structures of a random MD configuration 146
3.10　Radial distributions of the In atoms obtained from the MD and FP configurations 150
3.11　Phonon mode occupations obtained from linearly combining each of the FP and MD configurations 151

1.1　Quantum conductance spectrum for electron transport along the chain direction calculated for In/Si(111) model structures. 156
1.2　Quantum conductance spectra and isodensity surface of the local DOS for ideal and adatom-perturbed In nanowires. 157
1.3　Calculated and measured RA-spectra and band structure for In/Si(111)-(4×1)/(8×2) 158
1.4　Comparison of theoretical and experimental phonon spectra. Low energy shear, rotary and hexagon rotary modes, that facilitate the phase transition. 159
1.5　Radial distributions of the In chain atoms and phonon mode occupations obtained from linearly combining each of the FP and MD configurations 160
2.6　LEED and STM images of oxygen-doped and optically pumped In nanowires. 164
1.1　Calculated eigenvectors of phonon modes responsible for the phase transition. 167
1.2　Difference of the free energy $F(T)$ calculated for the (4×1) and (8×2) phase of the Si(111)-In nanowire array. 168
1.3　Phonon density of states calculated for the (4×1) and (8×2) phase of the Si(111)-In nanowire array 169

Concern for man and his fate must always form the chief interest of all technical endeavors. Never forget this in the midst of your diagrams and equations.

– Albert Einstein

Chapter 1
Introduction

1.1 Motivation

The classical wire as a single, usually cylindrical string of metal has a history that ranges back as far as the 2nd millennium BC in Ancient Egypt. At the beginning it was used for jewellery purposes only, but soon evolved into one of the most versatile and almost omnipresent construction materials of human history. Today the wire serves in an almost countless number of applications, as diverse as, i. e., architecture and construction, electric appliances, wire netting and fencing, protective clothing and even musical instruments. One of its most notable

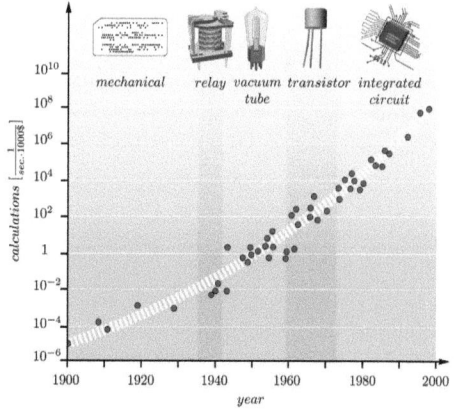

Figure 1.1: *Development of the speed increase per unit of cost for computing devices since 1900 [1, 2].*

uses is its employment by the electrical and electronic industry for telecommunications, power transmission, motors and generators, lighting, heating and many other purposes. The sizes of wires range from huge high voltage power transmission cables to tiny electronic device interconnects. With the advent of the microelectronics industry, the wire has seen a tendency to be scaled down to ever smaller dimensions, i. e. in chip bonding[1] and – most importantly – as *interconnects* on the chip itself.

---

[1] Nowadays the classical wire bonding technique of connecting the chip to the contacts of the carrier package is more and more replaced by the flip-chip technique that incorporates contact pads on both carrier and die instead of wires.

Beginning with the 11th U. S. Census in 1890, which was conducted employing Herman Hollerith's electric tabulating system for the first time, computing devices have consistently increased in speed and miniaturization (cf. Fig. 1.1). The continuing miniaturization, resulting ultimately in integrated circuits, has been quantified by Gordon Moore's famous law in 1965 [1]. As Moore's law holds even today, the number of transistors on a chip doubles roughly every 18 months with the commercially available minimum structure size currently bordering on 32nm. *This shrinking process represents a critical growth factor for a major part of the electronics and semiconductor industry.* However, the present silicon based devices are expected to reach their technological limit with respect to miniaturization in the not so distant future. Significantly smaller future *molecular electronics* device concepts have already been proven in the laboratory, such as, i. e., single molecule diodes [3] and amplifiers [4] and even single layer graphene transistors [5]. Naturally, to connect such molecular components in a useful device the interconnects need to be scaled down as well. The traditional very large scale integration (VLSI) interconnects consist of narrow copper-filled trenches etched into the substrate, that are then stacked in several layers (cf. Fig. 1.2). Unfortunately, for the future nanoelectronics this interconnect concept is not sufficiently scalable. Instead a paradigm change of the technology in use is required.

Today's microelectronics faces an especially pronounced transition: originally, the two grand theoretical frameworks developed in the 20th century – relativity and quantum mechanics – emerged from the desire to learn and understand the laws of nature. By now especially quantum mechanics is applied with tremendous success in present day's electronic devices. *However, as devices approach the atomistic scale, subtle quantum effects that arise from the microscopic size of the device itself, e. g. quantum interference or quantized electron transport, become more and more important even at room temperature.* The chemical nature of the metallic element starts to play an essential role. While, i. e., in the macroscopic world gold is a better conductor than lead by an order of magnitude, for conduction through a single atom, lead beats gold by a factor of three [8]. Even in today's 32 nm process the thinnest

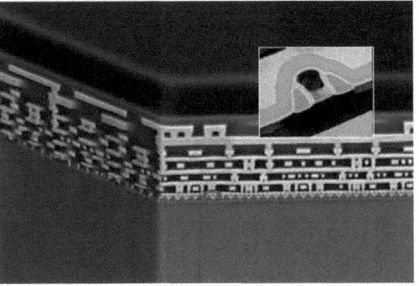

Figure 1.2: *Scanning electron microscopy cross section of a microprocessor in a 45 nm fabrication process. A transistor's gate insulating oxide layer (marked green) is 1.3 nm thick, corresponding to about 6 atomic layers. Image source: AMD*

1.1. Motivation

structure, the transistor's 1.2 nm thin gate insulating oxide layer, corresponds to no more than 6 atomic layers already (cf. Fig 1.2 inset). Thus quantum effects arising from the microscopic size of electronic devices will have to be considered in the near future. Designing such devices close to the atomistic scale poses a major challenge. Combined with the ability to manipulate atoms and molecules with increasing precision, this is currently driving a huge research effort into the properties of *atomic-scale nanowires*. Such nanowires may be used in future nanoelectronic devices as both interconnects and active circuit elements.

Regarding microscopic electron conduction the most notable consequence is that the well known Ohm's law looses its validity. At these dimensions electron transport occurs within the ballistic regime. Here the resistance does no longer originate from scattering processes within the conductor, but instead is caused by a finite reflection probability during the ejection of the electrons from the contact into the conductor. This reflection probability – or contact resistance – is caused by the confinement of the electrons in microscopic conductors. In comparison to the larger contacts the confinement induces a much stronger quantization of the electrons' wave-vectors perpendicular to their propagation direction. Thus the number of current carrying modes inside the conductor is much smaller than in the contacts. Hence the contact resistance arises from the necessary redistribution of the current among the modes at the contact-conductor interface. The resistance of a microscopic conductor is thus determined by its number of modes or, more accurately from a quantum mechanical point of view, by its *electronic structure*.

Nanowires are also very promising as active circuit elements that enable highly innovative nanoelectronic device concepts. The present day's microelectronics is based almost entirely upon the concept of the controlled creation of defects or impurity doping to design materials with specifically tailored electronic properties. Such doping effects are much more pronounced in low-dimensional systems due to the increased correlation between electrons. Doped nanowires might one day serve simultaneously as both interconnects and active logic circuit elements, i. e. as nanoscale transistors [9], representing a major step towards the ultimate limit of digital logic: In *reversible computing* as governed by the Landauer principle [13] the physical processes underlying the logic functions are (almost) reversible [14]. This is predicted to be the only possible way to improve the efficiency of computing beyond the fundamental von Neumann-Landauer limit of $kT \cdot \ln 2$ units of energy dispersed per irreversible bit operation. By avoiding the need to increase the information entropy – i. e. by "turning" a logic bit in a memory cell into heat by grounding the cell and by scattering losses – the amount of heat to be dissipated can be greatly reduced and significantly higher computational densities may thus be achieved [16, 17].

However, nanowires offer another very interesting prospect beyond the present day's electrical engineering: Today's applications are based on three fundamental types of devices: resistors, capacitors and inductors. In 1971 Chua discovered that these three device types do not represent a complete basis set in current-voltage (IV) space [10]. From symmetry arguments he deduced the existence and properties of a *fourth fundamental circuit element: the memristor*. While this work was largely ignored for more than 30 years, Williams et al. demonstrated in 2008 that memristance effects arise naturally at the nanometer scale [11, 12]. They constructed a memristive nanowire crossbar circuit (cf. Fig. 1.3) that features a wide range of possible applications, i. e. as a transistor replacement in non-amplifying applications such as neural networks, programmable logic, signal processing and as an ultra high density non-volatile memory.

Figure 1.3: *Atomic force microscopy image of a memristive nanowire crossbar circuit [11]. Image source: HP*

Thus the future necessity for a new type of VLSI interconnect and innovative nanoelectronic device concepts appearing on the horizon recently inspired a highly increasing interest in nanowire research. To understand and accurately predict the properties to be met upon further size reduction, the microscopic size of such nanowires requires a treatment by state-of-the-art quantum mechanical approaches.

Apart from the urgent technological needs described above, such one-dimensional (1D) electronic systems are also highly interesting from a fundamental point of view. Due to the increased electron correlation in low-dimensional systems, 1D structures feature a multitude of fascinating physical properties. Most notably, the commonly used Fermi-liquid model is no longer applicable. Instead this new state of matter exhibits radically different properties and has to be described in terms of the Luttinger-liquid model [19]. Among its hallmark features is the *spin-charge separation*. The elementary excitations of a Luttinger-liquid are *spin-density (SDW)* and *charge-density waves (CDW)* that propagate with different group velocities. They are thus completely unlike the quasiparticles of the Fermi-liquid, which always carry both spin and charge.

Even the question regarding the existence of a true 1D-electronic system is not a simple one. Peierls's theorem [20] states that a 1D-chain of atoms is unstable with respect to a periodic lattice distortion, leading to the formation of dimers. This dimerization opens band gaps at multiples of $k = \pi/a$, where a denotes the lattice parameter of the 1D-chain. The formerly half-filled band is now split into two bands, one filled and one empty. As a consequence, the electrons are slightly lowered in energy due to the distortion of the bands near the newly formed gap. Hence the system becomes insulating. This so-called *Peierls instability* can be seemingly circumvented by fixating the 1D-chain on top of a substrate. However, few such systems are known and most feature phase transitions towards lower temperatures due to the Peierls transition or charge density wave (CDW) formation [21].

For these fundamental and technological reasons 1D electronic systems are currently a subject of intensive investigation. One the few known systems are the atomic-size In-nanowires, that self-organize on the Si(111) surface [22]. This system is readily prepared experimentally and also accessible from a theoretical point of view. Additionally the In surface states are located almost completely inside the Si band gap. Hence this system is ideally suited as a testbed for the study of quasi-1D systems. The following section provides an introduction to the Si(111)-(4×1)/(8×2)-In nanowire substrate-adsorbate system presenting an overview of both previous research and today's open questions.

1.2 Previous Research

1.2.1 Structure and phase transition

While the Si(111)-(4×1)/(8×2)-In substrate-adsorbate system is currently under intensive investigation, it was originally discovered by Lander and Morrison as early as 1965 [22]. They examined surface reconstructions forming for various In coverages and temperatures between room temperature (RT) and several 100 °C. For a coverage of one In monolayer (ML) and subsequent tempering at 500 °C the formation of a (4×1) reconstruction was observed. However, it took another 30 years until an important and highly intriguing property of this reconstruction was discovered: In 1995 and

1997 direct [23] and indirect [24] photoemission experiments proved that the (4×1) reconstruction represents in fact a *quasi-1D metal* with three metallic In surface bands.

While early approaches to develop a structural model were unsuccessful, Bunk et al. [25] suggested in 1999 a model based on x-ray diffraction data that is consistent with the experimental observations. In this structural model – which is generally accepted today – the In atoms form two parallel zig-zag chains. These are separated by Si zig-zag chains resembling the π-bonded chains of the clean Si(111)-(2×1) surface (cf. Fig 2.4). The substrate itself remains unreconstructed, while the atoms constituting the uppermost layer are positioned at three different heights along the surface normal. The outer In atoms are positioned at the largest height, followed by the inner In atoms and finally the Si atoms at the lowest sites. This is consistent with *scanning tunneling microscopy* (STM) experiments [26] and *X-ray photoemission spectroscopy* (XPS) experiments, that indicate two different types of inequivalent In atoms as well [27]. Density functional theory (DFT) total energy calculations also confirm this structural model [29].

Shortly after a structural model for the (4×1) reconstruction was established Yeom et al. reported the observation of a temperature-induced reversible phase transition between (4×1) translational symmetry and a semiconducting (4×"2") reconstruction. The surface was probed by *reflection high energy electron diffraction* (RHEED), *angle-resolved photoemission spectroscopy* (ARPES) and STM techniques. Streaks found in the RHEED-patterns below a critical temperature of 120K indicate a period doubling of the surface reconstruction [30]. The emergence of streaks instead of dot-like reflections suggests the presence of some remaining disorder in the low temperature phase. STM experiments also reveal a corresponding modulation of the charge density. Fig. 2.5 shows high-

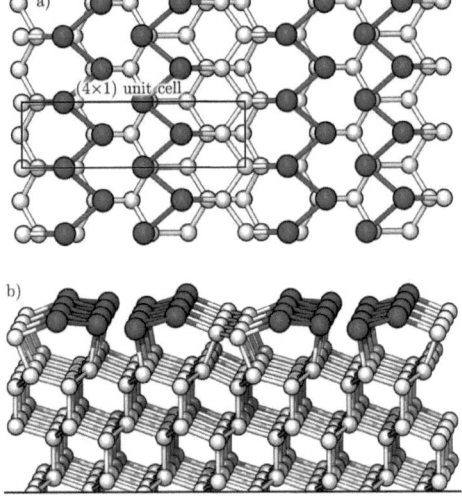

Figure 2.4: *Schematic top (a) and side views (b) of the (4×1) structural model for the In/Si(111)-(4×1) room temperature phase.*

1.2. Previous Research

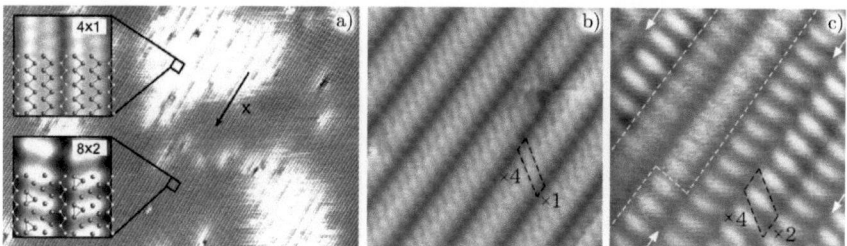

Figure 2.5: *a) Constant current STM images of the Si(111)-(4×1)/(8×2)-In surface obtained at 121K with high- and low-temperature phases coexisting in nanoscale domains [31]. The inset shows enlarged regions of the high- and low-temperature phases with superimposed structural models (cf. Figs. 2.4,2.6). b/c) An enlargement of a region that fluctuates between the two phases [30].*

and low-temperature phase STM images with the structural model of Fig. 2.4 superimposed. Since the STM data often exhibit an antiphase arrangement of adjacent (4×2) regions Yeom et al. concluded that the system's true ground state would be in fact a (8×2) reconstruction accompanied by a 1D charge density wave (CDW) locked in phase.

The ARPES spectra at 100K reveal an absence of states at the Fermi energy and thus imply the (4×"2") reconstruction to be semiconducting. Also according to the ARPES results the Fermi contours are "nested" with a nesting vector $2k_F = \pi/a_0$ ($a_0 = 3.84$) [30]. Thus Yeom et al. interpreted this phase transition as a Peierls transition. Simultaneously, they admitted that the unusually high ARPES band gap of 50-100 meV in relation to the thermal energy of about 10 meV at the critical temperature as well as the vanishing spectral weights of all three In surface bands at the Fermi energy could not be explained in terms of a simple Peierls transition.

In 2000 Kumpf et al. presented a x-ray diffraction investigation of the phase transition and reported the formation of a (8×2) reconstruction at 20K [32]. While the structure showed strong correlations between adjacent In chains, even at 20K the superstructure along the chain direction was not fully developed. However, a pure Peierls instability would have caused the CDW to condense into a fully developed superstructure at much higher temperatures for such a strong interchain coupling. This effectively rules out a simple Peierls transition as the phase transition's driving force. Based on their x-ray diffraction data Kumpf et al. derived a structural model for the (8×2) low temperature (LT) phase. As shown in Fig. 2.6a/b the In atoms in the (4×1) phase form trimers in antiphase arrangement between adjacent chains, leading to the

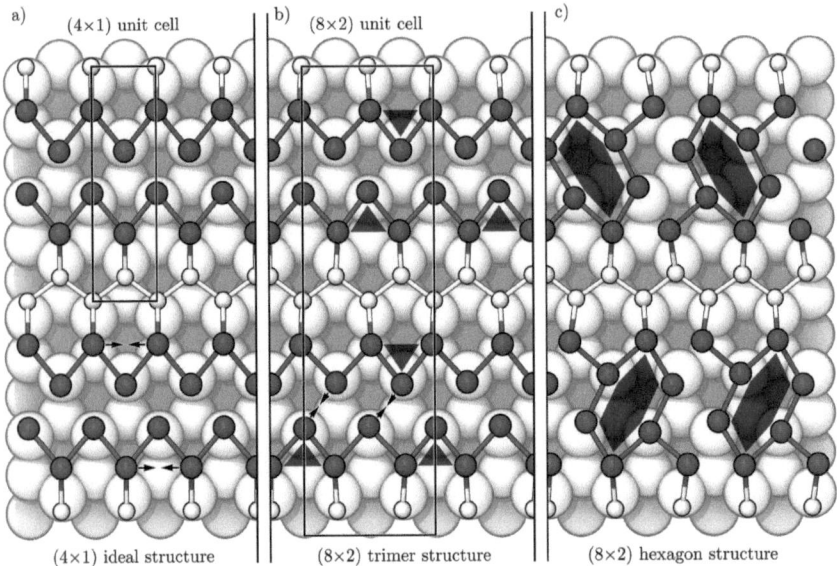

Figure 2.6: *Structural models for the low and high temperature phases of In/Si(111)-(4×1)/(8×2) as imaged in Fig. 2.5. Schematic top views of **a)** the ideal (4×1) model of the room temperature structure and **b/c)** the (8×2) trimer and hexamer structural models for the low temperature phase, respectively. Arrows indicate the movement of In atoms leading to the formation of trimers and hexamers.*

observed (8×2) reconstruction. Kumpf et al. also demonstrated that adjacent (4×2) subcells can be combined in several different ways to constitute a (8×2) unit cell. The streaky RHEED-patterns observed by Yeom et al. [30] can thus be explained in terms of "(4×2) subcell disorder", i. e. different arrangements of the (4×2) subcells along the chain direction. As the phase transitions driving force Kumpf et al. suggested either a triple-band Peierls instability or a reduction of the total free energy caused by the proposed In trimer formation and a subsequent relaxation of the top-layer Si atoms.

In 2001 Cho et al. presented a density functional theory (DFT) study [29] based on the (4×1) and (8×2) structural models suggested by Kumpf et al. Both the optimized atomic coordinates and the calculated electronic structure of the (4×1) structural model are in close agreement with the x-ray diffraction and ARPES data, respectively. Geometry optimizations in (4×2) and (8×2) unit cells employing the trimer model's coordinates as initial geometries also lead to optimized structures in close agreement with the structural models suggested by Bunk et al. It turns out that the

1.2. Previous Research

(4×2) structure is indeed more stable than the (4×1) structure by 8 meV per (4×1) unit cell. An antiphase arrangement of two (4×2) subcells in a (8×2) unit cell gains another 0.9 meV per (4×1) unit cell. This rather small difference in energy confirms the existence of different degenerate (8×2) ground states as assumed by Kumpf et al. Cho et al. concluded that the (4×2) subcells "freeze" randomly in one of these degenerate ground states in a temperature range between 20K and 100K. According to this model fluctuations above 100K give rise to a (4×1) reconstruction "on average".

However, while the geometric structure and total energy of the (8×2) trimer model is in close agreement with both experiment and theory, the electronic structure yielded a surprise: Cho et al. demonstrated that both the (4×2) and (8×2) trimer models still feature one band crossing the Fermi energy. *Thus they are still metals.* On the other hand, the ARPES data clearly show the LT phase to be semiconducting. Upon the exclusion of a misinterpretation of the ARPES data and DFT band gap problems one must conclude that the trimer model is not entirely correct. A more recent study of the temperature dependent conductivity by Uchihashi et al. also confirms the semiconducting state of the low temperature phase [33].

With increasing computational power performing DFT molecular dynamics (MD) simulations became numerically feasible for this system. Gonzáles et al. conducted a simulated annealing of the (4×1) reconstruction in a (4×2) unit cell. Upon cooling of the system they found a new structural model, that is more stable than the optimized trimer structure by 72 meV per (4×1) unit cell [34]. Graphically, this structure is obtained by a shear distortion of an In double chain by ±0.35 Å and a subsequent trimerization of the In atoms (cf. Fig. 2.6c). The resulting structure is termed *hexagon model* due to the hexagonal arrangement of the In atoms. Electronic structure calculations show this model to be semiconducting with a band gap of about 0.3 eV. This data also supports a driving mechanism for the phase transition, that was originally suggested by Ahn et al. [36]: The RT phase features three metallic bands with one exactly half-occupied and the remaining two occupied slightly less. During the phase transition a charge transfer is induced between the latter two bands, resulting in one empty and one now exactly half-occupied band. The two remaining metallic bands are now both exactly half-occupied and exhibit a Fermi contour nesting with the same nesting vector. This gives rise to a double band Peierls instability that opens a band gap and induces a period doubling of the unit cell. González et al. demonstrated that the shear distortion leads to the proposed charge transfer, while the trimerization subsequently opens a band gap. One year later – in 2006 – González et al. proposed another driving mechanism: Based on DFT-MD calculations they explained the (4×1) phase as a dynamical fluctuation between the four degenerate ground states of the (4×2) unit cell (two possibilities of trimerization for each of the two shear directions). MD calculations at 40 K show the atoms oscillating around the initial (4×2) hexagon structure. At 200 K the system fluctuates between the four degenerate ground states that González et al. identified as attractors. From this data they extrapolated the existence of a soft phonon mode inducing the shearing movement, that is the phase transition's driving mechanism according to this model. In a very recent study González et al. compare simulated and measured STM images [37]. Since they obtain the best agreement for sheared model structures they emphasize the importance of such a shearing mode again. However, whether such a soft shearing mode indeed exists remains unknown until today.

However, while indeed novel these results were received rather critically. The DFT code used by González et al. employs localized atomic orbitals as a basis set. While this is computationally very efficient, it also introduces a basis set dependence into the calculations because atomic orbitals do not constitute a *complete* basis set. Besides even qualitative differences of the electronic structure this may also affect the structural relaxation and total energy (discussed in detail in chapter 3.2, cf. present author's Publ. [14]). According to Cho et al. the (4×2) hexagon model is unstable within plane-wave DFT [38], which does employ a complete basis set. Given DFT's tendency to underestimate band gaps by a factor of 2 the gap obtained by González et al. is notably too large in comparison with experiment [42]. Yeom also commented that a driving mechanism by dynamic fluctuations is contradicted by the available photoemission data [39]. These photoemission experiments take place on a much faster time scale than the proposed dynamical fluctuations and consistently indicate a displacive-type phase transition between well-defined RT and LT structures. On the other hand, the hexagon structure is supported by recent positron spectroscopy data [40, 41].

In 2007 Fleischer et al. studied both the (4×1) and (8×2) reconstruction's surface phonons employing Raman spectroscopy [46]. At 250 K they identified 11 modes with A' and 1 mode with A'' symmetry. Cooling the system down to 60 K yielded 17 modes with A' and 8 modes with A'' symmetry[2]. All major modes of the (4×1) surface are also found in the (8×2) spectra, though blue-shifted and with increased

---

[2] The (4×1)/(8×2) surface features a *1m* (C_s) symmetry with exactly one mirror plane, located perpendicular to the In chains. A' and A'' denote modes whose corresponding atomic elongations are situated within and perpendicular to this mirror plane, respectively.

1.2. Previous Research

intensities due to the lower temperature. The new modes of the (8×2) surface are consistent with a backfolding of weakly dispersing phonon branches upon the phase transition induced doubling of the surface's unit cell. This was seen as supporting evidence for the Peierls model. Fleischer et al. also examined the measured spectra with respect to González et al.'s soft shear distortion model, where the (4×1) phase is assumed to be the time-average of dynamical fluctuations between degenerate (8×2) ground states. This model implies that phonon signatures of the LT phase should be observed also within the RT spectra. This is not the case effectively ruling out the dynamical fluctuation model. However, a soft shear mode that may be related to the phase transition in conjunction with a Peierls instability may provide an explanation for the single strong A″ mode along the chain direction at RT. While Bechstedt et al. had already calculated the (4×1) surface's phonon spectra in 2003 [47], a quantitative comparison between theory and experiment proved to be difficult. In Ref. [47] the topmost 4 atomic layers were taken into account leading to a multitude of calculated modes, while Raman spectroscopy probes the surface phonon modes only.

Reflectance anisotropy spectroscopy (RAS) is another technique that is highly sensitive to surface reconstructions and which is also widely applied in commercial applications. Since it involves the difference between spectra for two polarization axes it profits from a high degree of error cancellation and compares well with theoretically derived spectra. By comparing theory derived results with experimental data this technique allows to confirm or reject different structural models for surface reconstructions.

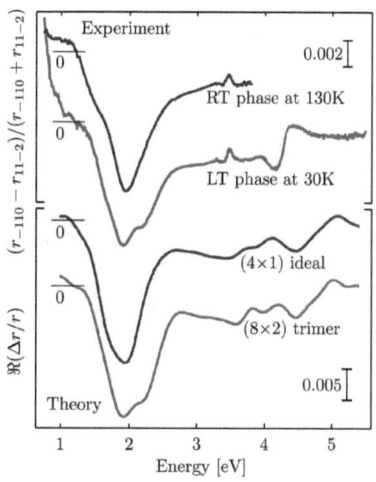

Figure 2.7: *Experimental vs. theoretical RA-spectra for the RT/LT phase and (4×1) ideal & (8×2) trimer models, respectively. Data is reproduced from Refs. [51, 52].*

Experimentally obtained RA-spectra at room temperature [49, 50, 51] exhibit an astonishing agreement with calculated spectra for the (4×1) structural model (cf. Fig. 2.7). Especially the strong anisotropy around 2eV is very well reproduced [52]. The experimentally observed splitting of the 2eV peak in conjunction with the (4×1)→(8×2) phase transition is consistent with the trimer model. However, no RAS has been calculated for the hexagon model so far. Additionally, the most interesting energy range lies within the mid-infrared (IR) regime, where the band structures and thus the spectra of different structural models differ the most. Neither measurements nor calculations have addressed the mid-IR regime due to both experimental and numerical difficulties. Since some of the competing structural models are metallic a meaningful RAS calculation needs to incorporate not only interband transitions, but *intraband* transitions as well. However, the impact on both the required computational time and memory is huge.

In conclusion, several different driving mechanisms for the phase transition have been suggested. Even after more than 10 years the matter is still intensively and controversely discussed. As both the available experimental and theoretical evidence is highly ambiguous neither the nature of the phase transition itself nor the associated ground state is clear. Total energy differences for competing structural models of the (8×2) ground state are within the numerical accuracy of ab initio calculations. This effectively disables any direct approaches to the LT phase's structural properties. Several measurements that could be used to distinguish between structural candidates, i. e. phonon spectra and RAS, lack the necessary theoretical data to be decisive.

1.2.2 Transport properties and doping

The discovery of the Si(111)-(4×1)-In surface's quasi-1D electronic character also sparked a growing interest for this system's transport properties, both from a fundamental and applied point of view. However, past studies of low-dimensional transport physics concentrated mainly on 2-dimensional electron gas systems (2DEG) formed at buried interfaces, such as GaAs/AlGaAs or metal-oxide Si interfaces [111, 112]. Measuring the conduction through surface-supported nanowires is even more challenging, because in this case the conductivity of interest σ_{ss} corresponds to the current carried by surface states. As the surface is always in contact with the substrate's bulk material, the measured conductance σ_{meas} includes also the bulk conductance σ_b and the conductance through a surface space-charge layer σ_{sc} arising from band bending effects beneath the surface. Thus the

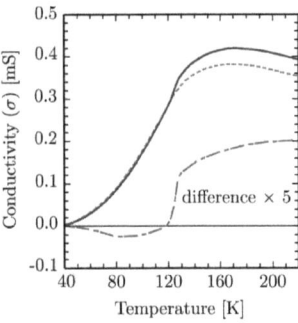

Figure 2.8: *Temperature dependence of the conductivity of a sample with intact (solid line) and severed (dashed line) In nanowires, respectively [33].*

1.2. Previous Research

measured conductivity accounts to $\sigma_{meas} = \sigma_{ss} + \sigma_b + \sigma_{sc}$ and needs to be corrected accordingly. Since even the electronic character of the LT phase was controversial (semiconducting/insulating vs. pseudogap opening) transport measurements were also expected to provide new insights regarding the nature of the phase transition.

In 2002 Uchihashi et al. measured the *macroscopic* transport properties of the Si(111)-(4×1)-In surface [33]. They deposited tantalum contact pads on the surface and first measured the conductance from pad to pad. Subsequently a large number of defects was introduced, effectively severing the nanowires. By measuring the conductance between the contact pads again the In surface contribution can be obtained as the difference between the two measurements. The conductance of the underlying bulk material and surface layer space-charge effects are thus excluded. Performing these measurements for different temperatures Uchihashi et al. obtained the results shown in Fig. 2.8. The conductance drops sharply at a transition temperature of 130K, indicating a metal-insulator (MI) transition around the same temperature as the previously observed structural phase transition.

Two years later in 2004 Tanikawa et al. reported novel transport measurements employing a four-point probe method with a microscopic probe spacing of $d = 8$ μm [44]. This method effectively confines the measuring current to the space-charge layer and surface states only, a priori excluding any bulk contributions without reverting to substraction techniques. They also demonstrated that below room temperature (RT) the measured conductivity is dominated by surface states, thus allowing for a *direct* measurement of the surface state conductivity for the first time. The temperature dependent conductivity $\sigma_{meas} = \sigma_{ss} + \sigma_{sc}$ is indicated by blue squares in Fig. 2.9, with the gray area representing the upper limit of the space-charge layer conductivity σ_{sc}. For temperatures below 200K it becomes

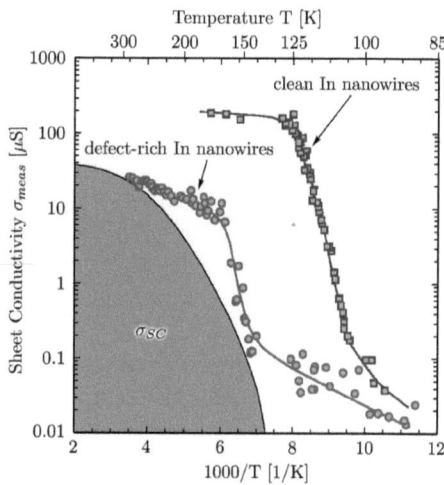

Figure 2.9: *Temperature dependent conductivity for clean and 0.1ML In-decorated In nanowires, respectively [44]. The upper limit of space-charge layer contributions is indicated by the shaded grey area.*

negligible, so that $\sigma_{meas} \approx \sigma_{ss}$. Below ~130K σ_{meas} features a steep decrease clearly indicating a metal-insulator (MI) transition (note the logarithmic scale). The measured band gap corresponds to a surprisingly large value of ~300meV. It is thus significantly larger than that estimated by PES measurements [45]. This may suggest a strong electron-phonon coupling accompanied by large fluctuations. From a theoretical point of view the impact of electron-phonon coupling upon the transport properties has not been addressed so far.

Interestingly, the MI transition occurs together solely with the formation of the (8×2) reconstruction. Any intermediate (4×"2") reconstructions between 200K and 130K – as observed by simultaneous RHEED measurements – were found to be metallic. These data represent strong evidence against a simple Peierls transition, where the (4×"2") reconstruction would already be non-metallic. This is also in accordance with the ARPES results by Yeom *et al.*, which predicted the (8×2) ground state to be semiconducting [30].

From a technological point of view the effect of impurities and defects on the wire conductance is also highly interesting, since current microelectronics is based almost entirely on the concept of tuning and modulating electronic device characteristics by the controlled creation of defects or impurity doping. Tanikawa *et al.* deposited an additional 0.1 monolayers (ML) of In upon the clean (4×1) reconstruction [44]. Performing the same conductivity measurements as for the clean surface they observed a significant decrease in conductivity by roughly one order of magnitude. The surface still features a MI transition, however, at a somewhat higher transition temperature of ~155K. Neither the conductance quenching mechanism nor the structural properties of this adatom decorated system are known.

In contrast an earlier STM and RHEED study in 2001 observed a reversion of the (8×2) low temperature (LT) phase to the (4×1) room temperature (RT) phase by depositing tiny amounts of In or Ag upon the (8×2) reconstruction at 100K [53]. While Ag reverted the (8×2) phase only locally around the adsorption site, In totally quenched the (8×2) symmetry reverting the whole structure back to the metallic (4×1) phase (cf. Fig. 2.10a). Thus the adsorption of adatoms on either the (8×2) LT or (4×1) RT phase clearly leads to completely different adsorbate systems.

Examining adatom species other than In and Ag – including H, O, Li, Na, Co, Pb – the reverse effect was discovered as well. Employing *low energy electron diffraction* (LEED),

1.2. Previous Research

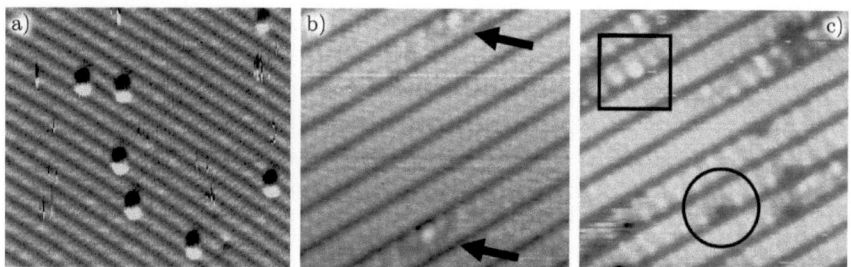

Figure 2.10: *a) Filled state STM image of the (8×2) LT phase after deposition of 0.1ML In [53]. The surface is reverted completely to (4×1) symmetry. b/c) Filled state STM images of the H and O adsorbed (4×1) RT phase, respectively [64]. Note the locally induced ×2 ordering of the In chains.*

high resolution electron energy loss spectroscopy (HREELS) and STM techniques Lee *et al.* observed a (4×1) → (4×2) metal-insulator phase transition at room temperature upon the deposition of an additional 0.2ML of Na [54]. Lee *et al.* interpreted their results in terms of the Na adatoms providing *pinning centers* to form a CDW upon deposition at RT as the phase transition's driving mechanism. However, a subsequent DFT study in 2002 by Cho *et al.* [55] found such a phase transition to be energetically unfavourable. Instead they found the Na adatoms to be highly mobile on the (4×1) surface and suggested the formation of finite 1D (4×2) Na wires on top of the (4×1) surface as the cause for the observed ×2 order. Electronic structure calculations for this structural model mostly explain the experimental data from Ref. [54].

In a very recent study from 2009 Shim *et al.* report a tuning of the (4×1)→(8×2) phase transition's critical temperature T_C also by depositing tiny amounts of Na [56]. They found T_C to decrease almost linearly from 126K down to 92K up to Na coverages of 0.015ML. Shim *et al.* proposed Na to act as an electron donor, causing an upward shift in the Fermi level. Within the CDW model the Fermi nesting vector increasingly deviates from its ideal condition for the commensurate CDW transition (cf. Fig. 2.11). Interestingly, a lowering of T_C can also be achieved by optical pumping where excess electrons are added to the system originating from the Si substrate [57].

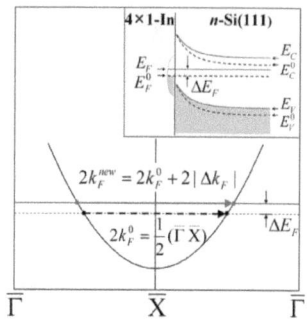

Figure 2.11: *Surface state dispersion and band bending (inset) before/after (dashed/solid lines) Na doping [56].*

To explain this observation a Fermi level shift and subsequent nesting vector deviation was adopted in this case as well.

Several studies report a (4×2) period doubling modulation for hydrogen adsorption similar to the one observed upon Na deposition, also with decreasing T_C (cf. Fig. 2.10b) [58, 59, 60, 64]. However, this H induced (4×2) phase seems to be fundamentally different from both the LT and Na induced (4×2) phases, since transport measurements show that H deposition barely affects the conductance [58].

While a decrease of T_C is readily achieved upon adsorption of various adatom species, O adsorption remarkably *increases* T_C. Other than in the previous cases the (4×1)→(8×2) phase transition is supported rather than hindered [64]. The clean (4×1) reconstruction's Fermi nesting vector is slightly smaller than in the perfect commensurate case [29]. Thus by oxygen hole-doping the Fermi level is lowered and as a consequence the nesting vector approaches the perfect Fermi-surface nesting condition for a commensurate CDW [64]. T_C might also be increased by a spatial correlation of the O defects that would then interfere constructively helping the CDW to condensate. So far neither spatially correlated nor mobile O defects have been observed experimentally, at least not before cooling. However, both Pb and Co defects have been reported to self-organize by substrate-mediated indirect interactions such as, i. e., Friedel oscillations [61, 63]. In case of Pb such indirect interaction effects are pronounced up to a range of 5 (4×1) lattice constants a, corresponding to a length of 19.3 along the In chains [61]. Depending on the exact defect placement of even or odd multiples of the lattice constant a a CDW condensation is either supported or inhibited, respectively.

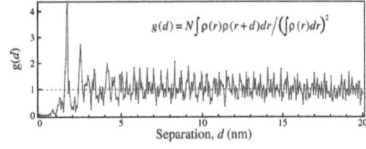

Figure 2.12: *Filled state STM image of the 0.02ML Co adsorbed (4×1) surface. The pair correlation function g(d) features a strong preference for Co-Co distances of 1.75nm and 2.5nm, respectively [63].*

In conclusion, the 1D transport properties of the Si(111)-(4×1)/(8×2)-In system have been measured for the ideal and In/H decorated surfaces. Impurity doping has been examined for several other adatom species as well. Doping itself represents the basic building block of nearly all present day microelectronics. As its effects are strongly enhanced in low-dimensional systems such studies are highly interesting with respect to future nanoelectronic device concepts.

However, no systematic study with regard to the structure of various defect types and their impact upon the system's transport properties is presently available. Despite its importance this system's transport properties and their temperature dependence as well as doping effects are barely understood today.

1.3 Project outline

The previous calculations and experiments presented above form a very solid foundation from which the ground state of the In nanowire array, its response to doping by adatoms, the impact of doping on the electron transport properties and the thermal and optical properties of this system can be explored. With respect to future nanoelectronic device concepts this project aims at generating a solid knowledge and detailed understanding about the physical foundations of such perspective future devices. It employs the prototypical Si(111)-(4×1)/(8×2)-In nanowire array as a suitable model system and contributes to the following fields of study:

1. What is the precise geometry and electronic structure of the In nanowire array's (8×2) LT phase? What mechanism drives the (4×1)→(8×2) phase transition?

2. What are the transport properties of the clean In nanowire array? How are they influenced by thermal excitations?

3. How does doping by adsorbed atoms modify the electron transport through the In nanowires? What is the precise mechanism of the conductance modification?

The present work employs state-of-the-art *ab initio* density-functional theory (DFT) calculations within both the generalized gradient and the local density approximation (GGA/LDA). Chapter 2 presents an introduction to the employed methodology for the derivation of structural, transport and optical properties.

It turns out that the DFT total-energy results for the In/Si(111)-(4×1)/(8×2) surface are extremely sensitive with respect to the details of the electron-electron interaction treatment. Thus the influence of semicore-electrons and pseudopotentials is checked thoroughly and systematically in chapter 3. Subsequently, the electronic and transport properties of the clean In/Si(111)-(4×1)/(8×2) surface are derived. The transport calculations are performed by means of a combined Green's function and localized orbital approach.

In chapter 4 these transport calculations are extended towards doped In nanowires. The influence of defects upon structure and conductance as well as the conductance modification mechanisms are discussed in detail and are compared with experiment to the extent of available measurements. The results allow for a tuning of the nanowires' conductance in a wide range by different mechanisms. In most instances experiments to confirm these predictions have yet to be performed.

In chapter 5 calculations of the anisotropic optical response in the visible and mid-infrared spectral range including intraband transitions are presented. It is demonstrated that states close to the Fermi energy lead to distinct optical fingerprints in the mid-infrared regime for each of the examined structural models. In comparison with a recent experiment these data are suitable to settle the 10-year old dispute about the In nanowire array's true LT ground-state.

Vibrational and thermal properties at finite temperatures as well as their influence on the In nanowire array's electron transport properties are examined in chapter 6. Large-scale frozen-phonon (FP) calculations and molecular dynamics (MD) simulations finally allow for the quantitative analysis of longstanding Raman spectroscopy measurements and an attempt to explain the driving mechanism of the phase transition. The existence of the soft shearing mode, as proposed by González *et al.*, together with two further soft modes facilitating the phase transition are confirmed for the first time. Furthermore, it is demonstrated that the employment of a classical energy distribution by MD simulations leads to inaccurate results for this highly subtle system. Several renowned earlier studies are affected and thus need to be revised. A more accurate scheme that incorporates both the correct energy distribution and potential energy landscape is proposed.

Finally the 7th and last chapter summarizes the main results of the present work and suggests some future research directions.

Today we reached a stage in physics that is different than anything in history before. We have a theory [...], so why do we not simply check whether it is true or false? Because to check it we need to calculate the consequences of this theory first. This time it is this first step that is the problem.

– Richard P. Feynman

CHAPTER 2

Methodology

2.1 Density Functional Theory (DFT)

The principal workhorse for the project outlined during the introduction is the framework of density functional theory (DFT). It is a calculational scheme to efficiently solve the Schrödinger equation for complex many-body systems from *first principles*. The term *first principles* – also known as *ab initio*, the latin term for *from the beginning* – implies a scheme based purely on the laws of quantum mechanics without requiring any empiric or semi-empiric parameters entering the calculation. This approach can thus be expected to yield highly accurate predictions regarding material properties in close agreement with experiment. The approaches regarding structural, thermal, transport and spectroscopic properties employed within the present work are based solely on these parameter-free DFT calculations.

2.1.1 The many body problem

Any material is comprised of atoms and molecules, which in turn consist of nuclei and electrons. Describing these materials thus implies correctly describing the mutual interaction of nuclei and electrons. In the non-relativistic case a quantum mechanical system's wavefunction $|\psi\rangle$ is governed by the Schrödinger equation

$$\mathcal{H}_{en}|\Psi\rangle = i\frac{\partial}{\partial t}|\Psi\rangle \tag{1.1}$$

with the Hamilton operator

$$\mathcal{H}_{en} = \mathcal{T}_e + \mathcal{T}_n + \mathcal{V}_{ee} + \mathcal{V}_{en} + \mathcal{V}_{nn} \tag{1.2}$$

Included in this Hamiltonian are the electron kinetic energy $\mathcal{T}_e$, the nuclear kinetic energy $\mathcal{T}_n$, the electron-electron interaction potential $\mathcal{V}_{ee}$, the electron-nuclear interaction potential $\mathcal{V}_{en}$ and the nuclear-nuclear interaction potential $\mathcal{V}_{nn}$. In case of an explicitly time-independent Hamiltonian $|\Psi\rangle$ can be expanded in terms of the Hamiltonian eigenfunctions $|\psi_i^{en}\rangle$ and eigenvalues E_i^{en}:

$$|\Psi(t)\rangle = \sum_i \langle \Psi(0)|\psi_i^{en}\rangle e^{-iE_i^{en}t}|\psi_i^{en}\rangle, \qquad (1.3)$$

$$\text{with } \mathcal{H}_{en}|\psi_i^{en}\rangle = E_i^{en}|\psi_i^{en}\rangle \qquad (1.4)$$

The Born-Oppenheimer approximation [66] allows to decouple the nuclear degrees of freedom $\{R_n\}$ from the electronic degrees of freedom $\{r_n\}$ and to neglect the coupling of electronic to nuclear motion due to the relatively "slow" motion of the nuclei in comparison to the electrons. Equation (1.4) is hereby separated into a nuclear and an electronic part:

$$\mathcal{H}_{en}(\{R_n\},\{r_n\}) = \underbrace{\mathcal{T}_n + \mathcal{V}_{nn}}_{\mathcal{H}_n(\{R_n\})} + \underbrace{\mathcal{T}_e + \mathcal{V}_{ee} + \mathcal{V}_{en}}_{\mathcal{H}_e(\{r_e\})} \qquad (1.5)$$

This allows for the separate solution of one Schrödinger equation for the nuclei and another one for the electrons. The wavefunction of the total system is then represented by a product-ansatz:

$$|\psi_i^{en}(\{R_n\},\{r_e\})\rangle = \psi_i^n(\{R_n\})|\psi_i^e(\{r_e\})\rangle \qquad (1.6)$$

The nuclear and electronic Schrödinger equations can thus be written as:

$$\mathcal{H}_n \psi_i^n(\{R_n\}) = E_i^n \psi_i^n(\{R_n\}) \qquad (1.7)$$
$$\mathcal{H}_e|\psi_i^e(\{r_e\})\rangle = E_i^e|\psi_i^e(\{r_e\})\rangle \qquad (1.8)$$

Throughout the present work the nuclei will be considered as classical particles, with the electrons coupled to the nuclei by Coulomb interaction only. The original quantum mechanical problem is hereby reduced to solving the Schrödinger equation for electrons moving in the electrostatic potential generated by the nuclei:

$$\mathcal{H}|\psi_i(\{r_e\})\rangle = E_i|\psi_i(\{r_e\})\rangle \qquad (1.9)$$

From here on the use of indices corresponding to the electrons/nuclei is discontinued, since only the electronic problem remains. It should be noted that both the electronic Hamiltonian as well as its associated wavefunction depend implicitly on the nuclear coordinates $\{R_n\}$. Assuming fixed nuclear positions the $\{R_n\}$ enter eq. (1.9) as parameters.

2.1. Density Functional Theory (DFT)

Bohr radius	$a_0 = 4\pi\epsilon_0\hbar^2/m_e e^2$	$0.529177 \cdot 10^{-10}$ m
Hartree energy	$H = m_e e^4/4\epsilon_0^2 h^2$	27.2114 eV
electron mass	m_e	$9.1 \cdot 10^{-31}$ kg
electron charge	e	$1.6022 \cdot 10^{-19}$ C

Table 1.1: *Atomic units used in the present work*

However, eq. (1.9) still represents a many-body problem. A multitude of different methods has been developed to solve this equation to the desired degree of accuracy. Most famous among these are the Hartree-Fock (HF) method and its further developments known as post-HF methods [67]. An extremely successful and widely used one is the framework of *density functional theory* (DFT). This method represents the building block of all the methods used throughout the present work and is thus examined in more detail in the following sections.

2.1.2 The Hohenberg-Kohn theorems

To solve the electronic problem in the external potential of fixed nuclei we consider a Hamiltonian of the following form:

$$\mathcal{H} = \mathcal{T} + \mathcal{V} + \mathcal{V}_{ext} \quad (1.10)$$

$\mathcal{T}$ represents the kinetic energy operator, $\mathcal{V}_{ext}$ the external potential including the potential caused by the nuclei and $\mathcal{V}$ the Coulomb interaction between the electrons. In atomic units (see Tab. 1.1) $\mathcal{V}$ can be written as:

$$\mathcal{V} = \frac{1}{|r - r'|} \quad (1.11)$$

This Hamiltonian is determined unambigously by the external potential and the total number of electrons N in the system. However, Hohenberg and Kohn [71] realized that by using the electron density $n(r)$ the problem of solving the Schrödinger equation can be formulated in a much more convenient way, while still being formally exact. The electron density is given as:

$$n(r) = N \int d^3r_2 \int d^3r_3 \ldots \int d^3r_N \psi^*(r_1, r_2, \ldots, r_N) \psi(r_1, r_2, \ldots, r_N) \quad (1.12)$$

Since the wavefunctions are normalized the total number of electrons N can be retrieved again from the electron density $n(r)$ by:

$$N = \int dr\, n(r) \tag{1.13}$$

Hohenberg and Kohn showed that eq. (1.12) is invertable, meaning that from a given *ground-state density* $n_0(r)$ it is possible to calculate the corresponding *ground-state* wavefunction $\psi_0(r_1, r_2, \ldots, r_N)$. While it seems impossible on first glance that a function of one variable r could be equivalent to a function of N variables $r_1...r_N$ the following theorems show that the knowledge of $n_0(r)$ implies much more than only the knowledge of an arbitrary function. These two theorems lie at the heart of DFT and are simply cited here without demonstration [71]. The first one legitimates the choice of n_0 as the central quantity over ψ.

Theorem I: *The external potential v_{ext} is an unambigous functional of the ground-state charge density n_0.*

This implies the charge density n_0 together with the total number of electrons N already determines the wavefunctions ψ unambigously. The total energy of the system can be expressed as a functional of the charge density according to

$$\begin{aligned} E[n(r)] &= T[n(r)] + V[n(r)] + V_{ext}[n(r)] & (1.14) \\ &= F_{HK}[n(r)] + \int dr\, v_{ext}(r) \cdot n(r) & (1.15) \end{aligned}$$

where $T[n(r)]$ and $V[n(r)]$ are the kinetic and potential energy of the electrons, respectively. $V_{ext}[n(r)]$ is the potential energy of the electrons due to the external potential. Furthermore Hohenberg and Kohn introduced the internal-energy functional $F_{HK}[n(r)]$ according to eq. (1.15). Since it does not depend on V_{ext} it thus allows to define the ground-state wavefunction ψ_0 as that antisymmetric N-particle wavefunction that minimizes $F_{HK}[n(r)]$ and reproduces n_0 [69].

The second theorem establishes the following variational principle:

Theorem II: *The density functional*

$$E[n] = F_{HK}[n] + \int dr\, v_{ext}(r) \cdot n(r) \tag{1.16}$$

assumes its minimum at the ground state density n_0.

These two theorems allow for changing from the use of the wavefunctions $\psi(r_i)$ to the density $n(r)$. Both schemes are formally equivalent and make no use of any approximations. Only the complexity of the equations to be solved is drastically reduced.

2.1. Density Functional Theory (DFT)

By calculating the wavefunctions directly $3N$ degrees of freedom would have to be taken into account, which is already computationally prohibitive for relatively small systems. Since $n(r)$ depends on only three coordinates it is much more convenient to use. Both theorems can be demonstrated for non-degenerate and degenerate ground-states, as well as for excited states. In the case of excited states $n(r)$ has to satisfy several more conditions. A detailed treatment can be found in Ref. [70].

2.1.3 The Kohn-Sham equations

While DFT can be implemented in many ways, the minimization of an explicit energy functional $F_{HK}[n]$ is usually not very efficient. A much more convenient way is the Kohn-Sham approach [72]. This approach is not based exclusively on the charge density but it also uses a special type of *single-particle* wavefunction. This allows the treatment of the Schrödinger equation as single-particle problem, while many-body effects are still included by the so-called exchange-correlation functional. First the functional $F_{HK}[n]$ is decomposed according to:

$$F_{HK}[n] = T_s[n] + V_H[n] + E_{XC}[n] \tag{1.17}$$

$T_s[n]$ represents the kinetic energy of a non-interacting electron gas and $V_H[n]$ its Hartree energy:

$$V_H[n] = \int\int dr\, dr'\, \frac{n(r)n(r')}{|r-r'|} \tag{1.18}$$

The remaining interacting terms $T - T_s$ and $V - V_H$ are merged into the exchange-correlation term E_{XC}. For non-interacting particles the total kinetic energy T_s equals the sum over the individual kinetic energies of each single particle. Therefore one can express $T_s[n]$ in terms of the single-particle wavefunctions of a non-interacting system with the density n according to:

$$T_s[n] = -\frac{\hbar}{2m}\sum_i \int d^3r\, \phi_i^*(r)\nabla^2\phi_i(r) \tag{1.19}$$

Since all ϕ_i are functionals of n, $T_s[n]$ is an explicit functional of the ϕ_i but unfortunately only an implicit functional of n with $T_s[n] = T_s[\{\phi_i[n]\}]$. Thus a direct minimization of $E[n] = F_{HK}[n] + V_{ext}$ with respect to the density n is impossible. To resolve this problem Kohn and Sham introduced a method that allows to conduct the necessary minimisation in an indirect way [72]:

$$\frac{\delta F_{HK}[n]}{\delta n(r)} \stackrel{!}{=} 0$$

$$= \frac{\delta T_s[n]}{\delta n(r)} + \frac{V_H[n]}{\delta n(r)} + \frac{E_{XC}[n]}{\delta n(r)} + \frac{V_{ext}}{\delta n(r)} \qquad (1.20)$$

$$= \frac{\delta T_s[n]}{\delta n(r)} + v_H(r) + v_{XC}(r) + v_{ext}(r)$$

Considering a system of non-interacting particles which move in an effective potential $v_s(r)$ the analogous type of minimisation can be done in a much simpler way because without interaction there are no Hartree and exchange-correlation terms:

$$\frac{E_H[n]}{\delta n(r)} = \frac{T_s[n]}{\delta n(r)} + \frac{V_s[n]}{\delta n(r)} = \frac{T_s[n]}{\delta n(r)} + v_s(r) \stackrel{!}{=} 0 \qquad (1.21)$$

This equation is solved by the density $n_s(r)$. From the comparison with eq. (1.20) it is clear that both minimisation problems yield the *identical* solution $n_s(r) \equiv n(r)$ if v_s is chosen as:

$$\boxed{v_s(r) = v_H(r) + v_{XC}(r) + v_{ext}(r)} \qquad (1.22)$$

So first Hohenberg and Kohn reduced the problem of having to solve the many-body Schrödinger equation to the problem of minimising the total energy functional $E[n]$ with respect to $n(r)$. Now Kohn and Sham again reduce the minimisation of $E[n]$ to solving the single-particle Schrödinger equation for a system of non-interacting particles moving in an effective potential $v_s(r)$. Thus the wavefunctions that satisfy the single-particle Schrödinger equation

$$\boxed{\left[-\frac{\hbar^2 \nabla^2}{2m} + v_s(r) \right] \phi_i(r) = \epsilon_i \phi_i(r)} \qquad (1.23)$$

of this replacement problem reproduce the *exact* density $n(r)$ of the original many-body problem,

$$\boxed{n(r) \equiv n_s(r) = \sum_i f_i |\phi_i(r)|^2} \qquad (1.24)$$

where f_i are the occupations of the wavefunctions ϕ_i. Eq. (1.22) to (1.24) are known as the Kohn-Sham (KS) equations. However, since v_H and v_{XC} depend on n, which in turn depends on the ϕ_i, which again depend on v_s, solving the KS equations represents a nonlinear problem. Therefore the problem is solved in an iterative *self-consistent* way. One starts with an initial guess for $n(r)$, calculates the corresponding $v_s(r)$ and solves Eq. (1.23) to obtain the ϕ_i. From these a new density is obtained by

2.1. Density Functional Theory (DFT)

Eq. (1.24) and the cycle is repeated until it converges. Once a converged solution has been found the total energy can be calculated according to

$$E_0 = \sum_i \epsilon_i - \frac{q^2}{2} \int d^3r \int d^3r' \frac{n_0(r)n_0(r')}{|r-r'|} - \int d^3r\, v_{XC}(r)n_0(r) + E_{XC}[n_0] \quad (1.25)$$

which is equivalent to Eq. (1.14). It should be noted that E_0 is not simply the sum over all ϵ_i, because the ϵ_i are not the eigenvalues of the original many-body problem, but of an auxiliary single-particle problem. They are thus completely artificial objects.[1] The N-particle wavefunction $|\psi\rangle$ also does not correspond to the wavefunction of the non-interacting Kohn-Sham system. Only the ground-state density n_0 and thus the total energy E_0 have a strict physical meaning and are formally exact.

However, the Kohn-Sham eigenvalues ϵ_i are still used to approximately calculate a large variety of physical properties. In many situations the ϵ_i provide a reasonable approximation.[2] This approach can be rationalised from the point of view of this mode of DFT as a *mean-field* theory (although with a very sophisticated mean-field $v_s(r)$). In fact most band-structure calculations in solid state physics are actually KS eigenvalue calculations. In practice this approach turned out to be highly successful, but not all resulting errors can be neglected. The KS eigenvalues underestimate the bandgap by up to 100% or erroneously even predict metallic behaviour for semiconducting systems. A variety of correctional methods are available for situations where the KS treatment of the many-body problem is insufficient. Among these are the GW-approximation (GWA) and Bethe-Salpeter equation (BSE), which will be treated later in chapter 2.3.

2.1.4 Exchange-correlation (XC) functionals

In principle the KS equations allow to calculate the ground-state energy and density exactly. However, the exchange-correlation term is unknown – although the HK theorems guarantee that it is a density functional – and has to be approximated for practical applications. Usually E_{XC} is decomposed into $E_{XC} = E_X + E_C$. While E_X can be expressed exactly at least in terms of the single-particle wavefunctions ϕ_i, no exact expression for E_C is known, neither in terms of the ϕ_i nor n. With the self-

---

[1] With the exception of the highest occupied KS eigenvalue, which corresponds to the ionisation energy of the system.
[2] These situations are typically characterised by the absence of fermionic quasi-particles and strong correlations.

interaction correction to the Hartree term V_H and the difference between the kinetic energy and its non-interacting counterpart E_{XC} contains all the many-body aspects of the problem:

$$E_{XC}[n] = (V[n] - V_H[n]) + (T[n] - T_s[n]) \qquad (1.26)$$

Thus the exact knowledge of E_{XC} would again require the exact solution of the many-body problem. Formally, the *adiabatic connection* approach allows to deduce two important properties of the exact XC functional, that in turn enable the derivation of reasonable approximations for E_{XC} [73]. These properties are only cited here. A demonstration can be found, i. e. in Ref. [75].

- E_{XC} depends only on the spherically averaged exchange-correlation density n_{XC}:

$$E_{XC}[n] = \frac{1}{2} \int dr\, n(r) \int_0^\infty dx\, 4\pi x n_{XC}^{sp}(r,x) \qquad (1.27)$$

$$\text{with } n_{XC}^{sp}(r,x) = \frac{1}{4\pi} \int_\Omega dr'\, \bar{n}_{XC}(r,r'),\ \Omega = |r-r'| = x \qquad (1.28)$$

As a consequence an approximation that reproduces the spherically averaged XC-density exactly, is sufficient to also calculate E_{XC} exactly.

- Integration of $\bar{n}_{XC}$ over r' yields the sum rule:

$$\int dr'\, \bar{n}_{XC} = \int dr'\, n(r')\bar{h}(r,r') = -1 \qquad (1.29)$$

Thus the XC interaction displaces exactly one electron from the environment of r. This phenomenon is the so-called XC-hole. The averaged density $\bar{n}(r,r')$, the pair-distribution function $\bar{h}(r,r')$ and thus the averaged XC-density $\bar{n}_{XC}(r,r')$ are given by:

$$\bar{n}(r,r') = n(r)n(r')(1+\bar{h}(r,r')) \qquad (1.30)$$
$$\bar{n}_{XC}(r,r') = n(r')\bar{h}(r,r') \qquad (1.31)$$

2.1.4a Local density approximation (LDA)

One of the practically most important types of approximations for E_{XC} is the *local density approximation* (LDA), which also finds extensive use in the present work. It was already suggested in the original work of Kohn and Sham [72] and – like almost all E_{XC} approximations – starts from the homogeneous electron gas. In the LDA E_{XC}

2.1. Density Functional Theory (DFT)

is approximated by integrating over the local contributions from the homogeneous electron gas' exchange-correlation density $\epsilon_{XC}^{hom}(n)$

$$E_{XC}[n] = \int d^3r\, n(r)\epsilon_{XC}[n] \approx \boxed{E_{XC}^{LDA}[n] := \int d^3r\, n(r)\epsilon_{XC}^{hom}(n)} \quad (1.32)$$

with $\epsilon_{XC}^{hom}(n) = \epsilon_X^{hom}(n) + \epsilon_C^{hom}(n)$. The corresponding exchange-correlation potential is then given by:

$$v_{XC}^{LDA}[n](r) = \left.\frac{\partial \epsilon_{XC}^{hom}(n)}{\partial n}\right|_{n \to n(r)} \quad (1.33)$$

The homogeneous electron gas was subject to extensive theoretical studies [70, 77, 78] and its *exchange* density is known exactly:

$$\epsilon_X^{hom}(n) = -\frac{3q^2}{4}\left(\frac{3}{\pi}\right)^{\frac{1}{3}} n^{\frac{1}{3}} \quad (1.34)$$

Thus E_X^{LDA} can be easily calculated:

$$E_X^{LDA}[n] = -\frac{3q^2}{4}\left(\frac{3}{\pi}\right)^{\frac{1}{3}} \int d^3r\, n(r)^{\frac{4}{3}} \quad (1.35)$$

Unfortunately the derivation of the *correlation* density is far more complicated because its determination for a homogeneous interacting electron gas is already a complicated many-body problem on its own. However, ϵ_C^{hom} can be calculated in some cases by perturbation theory [74, 75] or nowadays highly accurately by quantum Monte Carlo (QMC) approaches [76]. Subsequent fitting of the resulting values for ϵ_C^{hom} completes the construction of ϵ_{XC}^{hom} [79, 80].

Despite its simplicity LDA has proven surprisingly successful for many types of calculations, like band structures, structural properties or phonon modes, even when applied to systems which strongly differ from the reference system of the homogeneous electron gas. Typically, LDA systematically *underestimates* E_C but systematically *overestimates* E_X, resulting in a fortuitous error cancellation. The systematic character of this error cancellation is a consequence of the fact that the LDA XC-hole satisfies the sum rule in Eq. (1.29). LDA also gives a good approximation of the spherical average of the XC-hole since the LDA XC-hole exhibits an *a priori* circular symmetry. In Eq. (1.27) it was shown that only the spherical average of the XC-density contributes to E_{XC}. However, LDA works best for systems with slowly varying electron densities. Among its most serious shortcomings is its underestimation of bandgaps of up to

100%. It also tends to overestimate interatomic bond strengths resulting in slightly too small bond lenghts and too high cohesion energies.

2.1.4b Generalized gradient approximation (GGA)

While any real system has a spatially varying density $n(r)$, LDA exploits only knowledge of the density at the point r. Thus in many cases a systematic improvement of the LDA can be achieved by incorporating information about the gradient of the exchange-correlation density:

$$E_{XC}[n] \approx \boxed{E_{XC}^{GGA} := \int d^3r\, n(r)\epsilon_{XC}(n, \nabla n, \nabla^2 n, ...)} \qquad (1.36)$$

A large variety of these so-called *generalized gradient approximations* has been proposed over the last two decades. The most widely-used are the PW91 and PBE functionals, as suggested by Perdew, Wang [81, 82] and Perdew, Burke, Ernzerhof [83], respectively. Both types of GGA satisfy the sum rule in Eq. (1.29) and give improved results over LDA for a wide range of materials. Ground-state energies, molecular binding energies, hydrogen bonds and simple metal lattice constants are often described more accurately. However, GGA still has several fundamental shortcomings, such as its inability to describe dispersion-interaction[3], an underestimation of bond strengths, no discontinuity in $v_{XC}^{GGA}(r)$ with respect to N, causing an underestimation of bandgaps [85] and an unphysical exponential decay of the electrostatic potential above surfaces.

Both the LDA and GGA have been thoroughly tested with respect to their applicability in the present work, which will be described in more detail in chapter 3.

2.1.5 Periodic boundary conditions

In case of extended systems, such as crystals or surfaces, application of the Kohn-Sham formalism quickly becomes impractical due to the huge number of atoms to be treated. Furthermore the wavefunctions span an equally large space so that an excessive basis set would be required. Hence it is very useful to introduce periodic boundary conditions and thus employ the translational symmetry of the system.

---

[3] Dispersion-interaction is missing in LDA as well. Fortunately its incorporation is not required for the present work, but had to be accounted for in the water related projects mentioned in the summary (cf. chapter 7). This can be done, i. e., in a semi-empirical approach by the London dispersion formula [84]

2.1. Density Functional Theory (DFT)

2.1.5a Plane-wave basis set expansion

For periodic boundary conditions the external potential $v_{ext}(r)$ in which the electrons move is periodic with:

$$v_{ext}(r+R) = v_{ext}(r) \qquad (1.37)$$

The periodicity is then given by

$$R = \sum_i n_i a_i, \text{ with } n_i \in \mathbb{N}, \qquad (1.38)$$

$$\Omega = |a_1 \cdot a_2 \times a_3| \qquad (1.39)$$

with the a_i representing the lattice vectors of the primitive unit cell with the volume Ω. Introducing the reciprocal latticevectors $\{G\}$ by $R \cdot G = 2\pi n, n \in \mathbb{Z}$ allows the Fourier-representation of arbitrary lattice-periodic functions as:

$$f(r) = \frac{1}{\sqrt{\Omega}} \sum_G e^{iG \cdot r} f(G), \ f(G) = \frac{1}{\sqrt{\Omega}} \int_\Omega d^3r \, e^{-iG \cdot r} f(r) \qquad (1.40)$$

$\{G\}$ exhibits the same translational symmetry in reciprocal space as $\{R\}$ in real-space. In analogy to the Wigner-Seitz cell in real-space there is also a unit cell of highest symmetry in reciprocal space. This unit cell is the so-called Brillouin zone (BZ). For a periodic system of this kind any wave-vector q may be written as:

$$q = k + G, \ k \in BZ \qquad (1.41)$$

The translational symmetry of $v_{ext}(r)$ implies that the same symmetry applies to the Hamiltonian of this system. According to Bloch's theorem any eigenstate of such a lattice-periodic operator has to satisfy the condition [149]:

$$\phi_{nk}(r+R) = e^{ik \cdot R} \phi_{nk}(r) \qquad (1.42)$$

The quantum number n denotes the bandindex, which counts the various eigenstates at a given point k in the Brillouin zone. A naturally emerging type of basis set for this kind of translationally invariant system are plane-waves

$$\varphi_{kG}(r) := \langle r|kG \rangle = \frac{1}{\sqrt{\Omega}} e^{i(k+G) \cdot r} \qquad (1.43)$$

which satisfy the orthonomality- and completeness-relations

$$\langle kG|kG\rangle = \delta_{kk'}\delta_{GG'} \qquad (1.44)$$

$$\sum_{k,G}|kG\rangle\langle kG| = \hat{1}. \qquad (1.45)$$

As will be seen, this type of basis set is advantageous from both a physical and a technical point of view, with the electronic eigenfunctions / Bloch-states spread out over the entire system and a diagonal kinetic energy operator, respectively. In this basis set any electronic eigenstate can be expanded according to:

$$|\phi_{nk}\rangle = \sum_G c_{nk}(G)|kG\rangle, \quad c_{nk}(G) = \langle kG|\phi_{nk}\rangle \qquad (1.46)$$

Using the partitioning of the $\hat{1}$-operator according to Eq. (1.45) the Schrödinger equation of the Kohn-Sham system assumes the following form:

$$\sum_{k',G'}\left(-\frac{\hbar^2}{2m_e}\nabla + v_H[n](r) + v_{XC}(r) + v_{ext}(r)\right)|k'G'\rangle\langle k'G'|\phi_{nk}\rangle$$
$$= \sum_{k'G'}\epsilon_{nk}|k'G'\rangle\langle k'G'|\phi_{nk}\rangle \qquad (1.47)$$

All potentials in this equation are periodic with the lattice. With the help of Eq. (1.44), (1.46) and projection onto $\langle kG|$ the Kohn-Sham equation can then be transformed into the following numerically very convenient matrix representation:

$$\mathcal{H}_{G+k,G'+k}c_{nk}(G) = \epsilon_{nk}c_{nk}(G) \qquad (1.48)$$

$$\Rightarrow \sum_{G'}\left(\frac{\hbar^2}{2m_e}(k+G)^2\delta_{GG'} + v_H(G-G') + v_{XC}(G-G')\right.$$
$$\left. + v_{ext}(G-G')\right)c_{nk}(G') = \epsilon_{nk}c_{nk}(G) \qquad (1.49)$$

with $v(G-G') = \langle kG|v|kG'\rangle$

In this representation the kinetic energy becomes diagonal in reciprocal space. For the numerical evaluation the kinetic energy and the potentials have to be Fourier-transformed into reciprocal space. While the Hartree Term v_H may be transformed analytically using the Poisson equation, the other terms need to be transformed numerically by fast Fourier transformation (FFT) according to Eq. (1.40). The density

2.1. Density Functional Theory (DFT)

$n(G)$ results from the expansion coefficients $c_{nk}(G)$ as:

$$n(G) = \frac{2}{\sqrt{V}} \sum_{n,k}^{occupied} \sum_{G'} c_{nk}^*(G') c_{nk}(G' + G) \tag{1.50}$$

The expansion coefficients $c_{nk}(G)$ can be determined by diagonalizing the Hamiltonian $\mathcal{H}_{G+k,G'+k}$. In an actual numerical calculation the sum over G has to be truncated:

$$\frac{\hbar^2}{2m} |k + G|_{max}^2 \leq E_{cut} \tag{1.51}$$

Thus a single parameter E_{cut} controls the size of the plane-wave basis set. All of these plane-wave basis functions are orthogonal, ruling out any basis set superposition errors. The number of plane-waves N_{pw} for a given cutoff energy E_{cut} corresponds approximately to:

$$N_{pw} \approx \frac{\Omega}{6\pi^2} \left(\frac{2m_e}{\hbar^2}\right)^{\frac{3}{2}} E_{cut}^{\frac{3}{2}} \approx 2.27 \cdot 10^{-3} \; E_{cut}[eV]^{\frac{3}{2}} \; \Omega[\text{Å}^3] \tag{1.52}$$

2.1.5b Supercell method

The approach described in the previous section is only applicable for systems which are translationally invariant in all three spatial directions. This is naturally the case for crystals but not for lower dimensional systems like surfaces or molecules. Therefore an artificial periodicity is introduced by the so-called *supercell* approach, where the system is periodically repeated in all three dimensions, even if it has no actual translational symmetry in these directions. Hence, i. e., a surface will be modeled by a three dimensional superstructure consisting of alternating slabs of bulk material terminated by the desired surface and vacuum regions, respectively. The vacuum region must then be chosen large enough to ensure there is no relevant interaction between adjacent periodic images of the surfaces. Optionally one can also correct for surface-induced dipole moments by applying a suitable sawtooth-potential. The slab itself is either constructed in a symmetric way, where both sides are terminated by the desired surface and the middle-layer is kept fixed at the bulk lattice constant (see Fig. 1.1a). It can also be constructed asymmetrically by keeping one surface fixed at the bulk lattice constant and saturating the dangling bonds with hydrogen to simulate continued bulk material (see Fig. 1.1b). Both methods are applied in the present work, depending on the actual problem.

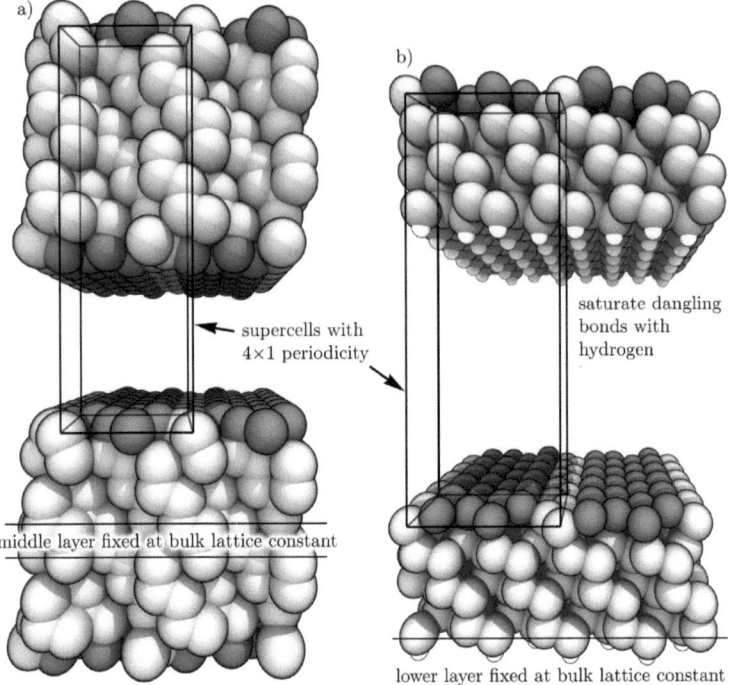

Figure 1.1: *Supercell methods for a) symmetric and b) asymmetric slabs. The representation is schematic and dimensions are not true to scale.*

An analagous method applies to molecular structures, where the supercell must again contain sufficient vacuum to minimize the interaction between the molecule and its neighbouring periodic images.

2.1.5c k-space integration

Calculating expectation values in many instances requires performing integrations over the Brillouin zone. I. e. in a bulk crystal all occupied states n contribute to the density $n(r)$ and thus to the potential $v_{ext}(r)$ at each point $k \in$ BZ. For numerical evaluation the number of k-points must be restricted and thus the integral over the Brillouin zone is reduced to a sum over the k-points. Considering a lattice-periodic function $f(k)$ the goal is to approximately calculate the integral

$$\bar{f} = \frac{1}{\Omega} \int_{\text{BZ}} d^3k \ f(k) \tag{1.53}$$

2.1. Density Functional Theory (DFT)

Therefore a way has to be devised to find a set of k-points that satisfies the requirements of a given accuracy at the least possible number of sampling points, while accounting for the symmetry of $f(k)$. A single point – the *mean value point* – at which the function value of the integrand approximates the mean value rather well was first identified by Baldereschi for the cubic lattice [86] and was later generalized for other lattices as well [87]. Chadi and Cohen [88] derived a method to calculate larger sets of special k-points, which was later also generalized for other types of crystal lattices [89]. According to their method the mean value $\bar{f}$ is calculated as a weighted sum over $f(k)$:

$$\bar{f} \approx \sum_{k}^{N_k} \omega_k f(k), \text{ with } \sum_{k} \omega_k = 1 \qquad (1.54)$$

The error of this approximation may be arbitrarily reduced by constructing increasingly larger sets of k-points. Today the most commonly used approach was suggested by Monkhorst and Pack, which contrary to the earlier methods, does not depend on the crystal lattice type [90]. The points of a Monkhorst-Pack (MP) mesh represent a mesh of $\prod_i q_i$ equidistant points, situated along the reciprocal basis vectors b_i of the unit cell. The mesh itself is centered at the Γ-point of the Brillouin zone.

$$k_{ijk} = u_i b_1 + u_j b_2 + u_k b_3 \qquad (1.55)$$

$$\text{with } u_i = \frac{1}{2q_i}(2r - q_i - 1), \; r = 1, 2, ..., q_i \qquad (1.56)$$

In this case the weights ω_k in Eq. (1.54) are constant, with $\omega_k = 1/\prod_i q_i$. By using symmetry properties one can reduce the summation to wave-vectors inside the irreducible wedge of the Brillouin zone only. The k-point weights ω_k have to be adjusted accordingly then. This scheme offers the advantage of the k-points being very easy to construct, while the accuracy can be arbitrarily increased by choosing sufficiently large q_i.

The Monkhorst-Pack scheme has proven very successful for semiconductors and insulators, but on first glimpse seems to fail for metals because at $T = 0$ K the function to be integrated becomes discontinuous at the Fermi energy. This problem can be resolved by introducing fractional occupation numbers into the evaluation of the charge density:

$$n(r) = \sum_{n,k} \omega_k f_{nk}^{occ} |\phi_{nk}|^2 \qquad (1.57)$$

The simplest approach is using the Fermi-Dirac distribution or Gaussian functions to calculate the occupation numbers f_{nk}^{occ}. Several other types of smearing functions have also been suggested over the years, i. e. by Methfessel and Paxton [91] or Marzari and Vanderbilt [92]. However, in the present work simple Gaussian smearing turned out to be well suited in most cases.

2.1.6 The pseudopotential approach

The number of basis functions scales with $\mathcal{O}(E_{cut}^{\frac{3}{2}})$ and an actual DFT calculation with about $\mathcal{O}(E_{cut}^{3})$. Hence the basis set expansion should be truncated as low as possible. The inter-atomic part of the wavefunction is usually well described by few plane-waves because there the KS wavefunctions do not vary very much. To the contrary, near the nuclei the KS wavefunctions feature much stronger fluctuations and oscillations which require a large number of plane-waves for the expansion series to converge. While computer technology is evolving quickly, the cutoff energy required for all-electron calculations still induces computational demands that renders large-scale calculations prohibitive. However, most of the interesting physics and chemistry take place in the overlap regions of the valence orbitals. Thus the core electrons contribute little of significance to properties like, i. e. the material's geometric structure, phonon frequencies and transport properties. A sensible approach is therefore to further reduce the electronic problem by the so-called *frozen-core approximation*.

2.1.6a Frozen-core approximation

In the frozen-core approximation the core electron wavefunctions are not anymore determined self-consistently. Instead they are taken from an all-electron calculation for the respective atomic species and will then be kept "frozen". That way the Kohn-Sham equations only have to be solved for the valence electrons, however, with a modified effective potential $v_s(r, [n_{val}])$:

$$\left[-\frac{\hbar^2 \nabla^2}{2m} + v_s(r, [n_{val}]) \right] \phi_i(r) = \epsilon_i \phi_i(r) \tag{1.58}$$

The effective potential $v_s(r, [n_{val}])$ only depends on the valence electron density $n_{val}(r)$, while the core electron density $n_{core}(r)$ remains unchanged:

$$\begin{aligned} v_s(r, [n_{val}]) &= v_{core}(r) + v_H(r, [n_{val}]) + v_{XC}(r, [n_{val}]) & (1.59) \\ v_{core}(r) &= v(r) + v_H(r, [n_{core}]) + v_{XC}(r, [n_{core} + n_{val}]) - v_{XC}(r, [n_{val}]) & (1.60) \end{aligned}$$

2.1. Density Functional Theory (DFT)

The effective core potential is comprised of the potential of the bare nuclei $v(r)$, the Hartree term for the core electrons $v_H(r, [n_{core}])$ and the core's exchange-correlation potential. In a formally exact way this XC-term can only be written as the difference between the XC-potentials with and without the core electron density, respectively. Thus it needs to be approximated:

$$v_{XC}[n_{core} + n_{val}] - v_{XC}[n_{val}] \approx \sum_j v_{XC}[n_{core}^{j,AE} + n_{val}^{j,AE}] - v_{XC}[n_{val}^{j,AE}] \quad (1.61)$$

$n_{core}^{j,AE}$ and $n_{val}^{j,AE}$ represent the core and valence electron densities of the atom j, with the summation ranging across all atoms contained in the system. They are obtained by performing all-electron (AE) calculations for the free atom j.

2.1.6b Pseudopotential concept

In the frozen-core approximation only the valence electronic problem remains. However, the shape of the valence wavefunctions remains unchanged. Thus oscillations of the valence wavefunctions near the nuclei may still affect the required cutoff energy and convergence behaviour. Extending the frozen-core approximation it may be assumed that not only the *core electrons* but also the *shape of the valence wavefunctions* is irrelevant for the physical and chemical properties of interest. For this purpose the effective core potential $v_{core}(r)$ is modified in a way that it yields the desired valence states [93]. Considering an atomic single-particle Schrödinger equation

$$\mathcal{H}|\phi_i^\alpha\rangle = \epsilon_i^\alpha |\phi_i^\alpha\rangle, \text{ with } \alpha = \{val, core\} \quad (1.62)$$

with the core and valence states $|\phi_i^{core}\rangle$ and $|\phi_i^{val}\rangle$, respectively, one can construct a so-called *pseudo valence state* $|\varphi_i^{val}\rangle$ by summing across core states:

$$|\varphi_i^{val}\rangle = |\phi_i^{val}\rangle + \sum_j a_{ij} |\phi_j^{core}\rangle \quad (1.63)$$

These pseudo valence states may be constructed in a particular way, that they do not exhibit oscillations near the core anymore. They satisfy a Schrödinger equation to the same eigenvalue as $|\phi_i^{val}\rangle$, but with a different potential v_{PS}:

$$(H + v_{PS})|\varphi_i^\alpha\rangle = \epsilon_i^\alpha |\varphi_i^\alpha\rangle, \text{ with } v_{PS} = \sum_j (\epsilon_i^{val} - \epsilon_j^{core}) |\phi_j^{core}\rangle\langle\phi_j^{core}| \quad (1.64)$$

The potential v_{PS} is called a *pseudopotential*. Introducing a suitable v_{PS} (which requires solving the all-electron problem) allows a dramatic reduction of the required cutoff energy and number of electrons to be treated explicitly, without any significant loss of accuracy. Due to certain conditions the v_{PS} have to satisy, many schemes exist for constructing suitable pseudopotentials.

2.1.6c Norm-conserving pseudopotentials (NC-PP)

A first approach to construct pseudopotentials usable in practical calulations was suggested by Bachelet, Hamann and Schlüter [94, 95] and was extended later by Troullier and Martins [96]. First the exact core and valence wavefunctions are obtained by solving the all-electron problem for a single isolated atom. A radial symmetric effective potential $v_s(\mathbf{r}) = v_s(r)$ is assumed which allows to separate the radial part of the all-electron Kohn-Sham equation. Thus the exact KS core and valence wavefunctions are known for reference. Subsequently a *core radius* r_{core} is chosen within which the all-electron wavefunction is to be replaced by a pseudo wavefunction (see Fig. 1.2). Naturally, r_{core} must be small enough to exclude any overlaps for the materials to be described. It should not be too small either, since a smaller r_{core} requires a larger basis set. Starting from a desired analytical wavefunction within r_{core} the inverse radial Schrödinger equation is solved for each angular momentum component l under the following constraints:

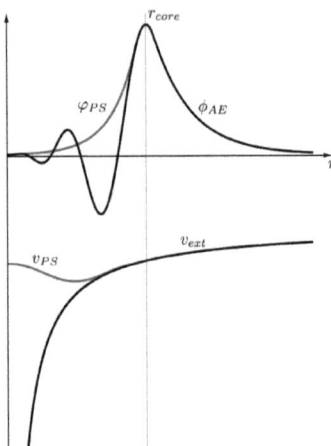

Figure 1.2: *Schematic illustration of all-electron and pseudo wavefunctions and potentials.*

- Normconservation: $\langle \varphi_i | \varphi_i \rangle = 1$

- Transferability: Equal atomic pseudo and all-electron eigenvalues $\epsilon_l^{PP} = \epsilon_l^{AE}$, with equal wavefunctions for $r > r_{core}$

- Softness: cusp- and node-free pseudo wavefunction for $r < r_{core}$

- Smoothness: pseudo and all-electron wavefunction are equal up to their fourth derivative at $r = r_{core}$

2.1.6d Ultrasoft pseudopotentials (US-PP)

Especially first-row elements like oxygen or carbon are ill-suited for a Troullier-Martin type pseudopotential description. The cutoff radius r_{core} needs to be very small, since

2.1. Density Functional Theory (DFT)

their $2p$ all-electron wavefunctions are strongly localized near the core. Hence the resulting pseudopotential is not very "soft" anymore and still requires a high cutoff energy with respect to the all-electron description. Based on a generalization of the Troullier-Martins scheme Vanderbilt suggested the concept of *ultrasoft* pseudopotentials with the following conditions [98]:

- All-electron and pseudo wavefunction do no longer need to exhibit identical scattering properties at eigenvalues ϵ_l^{AE}, but several arbitrarily chosen energies $\tilde{\epsilon}_{il}$. These $\tilde{\epsilon}_{il}$ can be conveniently placed at the energetic range of the physical characteristics of interest

- Remove restriction to normconserving pseudopotentials

- Relax smoothness condition from Troullier-Martin scheme

While the application of the second condition turns the Kohn-Sham equation into a generalized eigenvalue problem, the additional computational cost is usually compensated by the achieved basis set reduction. Within the ultrasoft approach the KS equation can be expressed in terms of the overlap operator $\mathcal{S}$ between the pseudo wavefunctions:

$$\mathcal{S} = \infty + \sum_{i,j} \mathcal{Q}_{ij} |\beta_i\rangle \langle \beta_j| \qquad (1.65)$$

$$\mathcal{Q}_{ij} = \langle \phi_i | \phi_j \rangle - \langle \varphi_i | \varphi_j \rangle \qquad (1.66)$$

$$|\beta_i\rangle = \sum_j \mathcal{B}_{ij}^{-1} |\gamma_j\rangle \qquad (1.67)$$

$$\mathcal{B}_{ij} = \langle \varphi_i | \gamma_j \rangle \qquad (1.68)$$

$$|\gamma_j\rangle = (\tilde{\epsilon}_j - \mathcal{T}_e - v_{loc}) |\varphi_i\rangle \qquad (1.69)$$

$$\text{with } \langle \varphi_i | \mathcal{S} | \varphi_j \rangle = \delta_{ij} \qquad (1.70)$$

v_{loc} represents the local part of the pseudopotential, with $v_{loc}(r) = v_{AE}(r)$ for $r > r_{core}^{loc}$. The KS equation then can be written in terms of $\mathcal{S}$ as:

$$\left[-\frac{1}{2}\nabla^2 + v_{PS} + v_H[n_{val}] + v_{XC}[n_{val}] + \sum_{i,j} \mathcal{D}_{ij} |\beta_i\rangle \langle \beta_j| \right] |\phi_i\rangle = \epsilon_i \mathcal{S} |\phi_i\rangle \qquad (1.71)$$

$$\mathcal{D}_{ij} = \int d^3r (v_H[n_{val}] + v_{XC}[n_{val}]) \mathcal{Q}_{ij}(r) \qquad (1.72)$$

Unfortunately the advantages described above are not without cost: Optical and spectroscopic properties do not compare well to the experiment, since the all-electron wavefunction is not retained.

2.1.6e Projector-augmented wave pseudopotentials (PAW)

The overlap operator S is the central quantity of the ultrasoft pseudopotential approach. Realizing that S can be expressed in terms of a transformation operator T, with $S = T^{\dagger}T$, represents the main step towards the projector-augmented wave approach by Blöchl [99]. The PAW approach is based on T instead of S and leads to an exact all-electron description within the frozen-core approximation. At the same time the computational cost is comparable to that of the ultrasoft scheme. By using radial projector functions one can transform between the all-electron wavefunctions defined on a radial grid and pseudo wavefunctions in a plane-wave basis set expansion. This way the original all-electron wavefunction is retained within the calculation without inducing excessive computational demands due to oscillations near the core. Any observable is hence evaluated as usual on the plane-wave grid. Near the core it is additionally evaluated on the radial grid using the all-electron wavefunction. Especially optical and spectroscopic properties profit significantly from this approach.

2.1.7 Calculation of the forces: Hellmann-Feynman theorem

Within the DFT framework presented so far it is possible to calculate the electronic structure of atoms, molecules and solids. It provides access to both the systems ground-state energy and Kohn-Sham type wavefunction. However, a description of the forces and thus a dynamic description of the atomic positions and equilibrium geometry is still required. In analogy to the expression $F(r) = -\nabla_r V(r)$ from classical mechanics, the quantum mechanical forces can be expressed as:

$$F = -\nabla_r \langle E \rangle, \text{ with } \langle E \rangle = \min\langle \phi | \mathcal{H} | \phi \rangle, \langle \phi | \phi \rangle = 1 \qquad (1.73)$$

The Hellmann-Feynman theorem allows to calculate the forces directly from the ground-state wave functions [101]. It states that for an *exact* eigenstate ϕ the following identity is valid for any degree of freedom λ, in this case the atomic coordinates:

$$\frac{\partial E}{\partial \lambda} = \underbrace{\langle \frac{\partial \phi}{\partial \lambda} | \mathcal{H} | \phi \rangle}_{=E\langle \phi |} + \langle \phi | \frac{\partial \mathcal{H}}{\partial \lambda} | \phi \rangle + \underbrace{\langle \phi | \mathcal{H} | \frac{\partial \phi}{\partial \lambda} \rangle}_{E\langle \phi |} \qquad (1.74)$$

$$= \langle \phi | \frac{\partial \mathcal{H}}{\partial \lambda} | \phi \rangle + E \frac{\partial}{\partial \lambda} \langle \phi | \phi \rangle = \langle \phi | \frac{\partial \mathcal{H}}{\partial \lambda} | \phi \rangle \qquad (1.75)$$

However, due to the variational nature of its calculation ϕ is expanded in a finite plane-wave basis set and is thus not an exact eigenstate. Hence the step from Eq. (1.74) to Eq. (1.75) is no longer valid, since the first and last term of Eq. (1.74) have

2.1. Density Functional Theory (DFT)

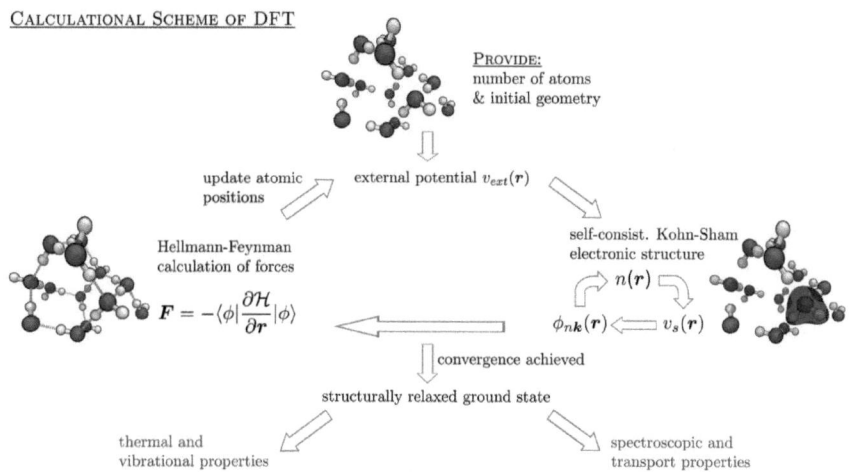

Figure 1.3: *DFT calculational scheme for deriving ground-state characteristics for a given system.*

to be considered explicitly. They only vanish if the basis set could be constructed in such a way that ϕ does not depend on λ, which is known as the Hurley condition [102, 103]. Fortunately the plane-wave basis set already satisfies this condition so that the forces may be obtained directly from the ground-state wavefunctions by means of the Hellmann-Feynman theorem:

$$\boxed{F(r) = -\langle \phi | \frac{\partial \mathcal{H}}{\partial r} | \phi \rangle} \quad (1.76)$$

A brief overview of the DFT calculational scheme is shown in Fig. 1.3. Starting with an initial user-supplied atomic geometry the external potential v_{ext} is constructed, possibly with the help of pseudopotentials. Subsequently the Kohn-Sham equations are solved in the *electronic* self-consistent loop. After reaching electronic convergence the forces can be calculated using the Hellmann-Feynman theorem. Then the atomic positions are updated according to the Hellmann-Feynman forces and the outer *ionic* loop is closed by recalculating the external potential. How electron transport and optical properties are obtained from the resulting structurally relaxed ground-state is the topic of the next sections.

2.2 Transport theory

This section begins with an introduction to the underlying concepts of transport physics and then moves on to the derivation of the Landauer formalism, that allows to evaluate a given systems conductance from a microscopic point of view in terms of a transmission probability. Subsequently the further treatment elaborates on the techniques required to implement the Landauer scheme on top of *first principles* electronic structure calculations.

2.2.1 Preliminary concepts

Considering semiconductors, electronic conduction may occur either through electrons in the conduction band or holes in the valence band. However, the present work is centered on *electron* transport, which is also the case in most experimental studies concerning micro- and mesoscopic conduction phenomena. Hence the following treatment is based on that very assumption. The dynamics of electrons in the conduction band can be modeled according to the effective mass equation [106]

$$\left[E_c(r) + \frac{(i\hbar\nabla + eA(r))^2}{2m} + \mathcal{V}(r)\right]\psi(r) = E\psi(r) \qquad (2.77)$$

where $\mathcal{V}(r)$ denotes the potential energy, $A(r)$ represents the magnetic vector potential, m the effective mass and $E_c(r)$ the conduction band edge energy. The wave-functions satisfying this equation have the form of plane-waves due to the implicit incorporation of the lattice potential by the inclusion of its effect into the effective mass:

$$\psi(r) = e^{ik\cdot r} \qquad (2.78)$$

While this description is a very simple one it allows to introduce several important concepts that are required for the derivation of the Landauer formalism in section 2.2.2. Hence one should be aware of the exemplary character here. However, the Landauer formalism itself and the later treatment in terms of Green's and Wannier functions is no longer based on this simple model, of course.

2.2.1a Subbands or transverse modes

Consider a conductor where electron transport is restricted in at least one dimension, i. e., a 2-dimensional electron gas (2-DEG)[4] or a nanowire where transport is

---
[4] A well-known experimental realization takes place at the interface layers in GaAs-AlGaAs heterostructures [111, 112].

2.2. Transport theory

unrestricted only along a single axis. Specifically for a two dimensional rectangular conductor that is uniform along x-direction and has a confining potential $\mathcal{V}(y)$, such as a nanowire array on a surface, the effective mass equation reads [106]:

$$\left[E_s + \frac{(i\hbar \nabla + eA)^2}{2m} + \mathcal{V}(y) \right] \psi(x,y) = E\psi(x,y) \tag{2.79}$$

Assuming a constant magnetic field B_z along the normal direction of the plane, with

$$A = -\hat{x} B_z y \Rightarrow A_x = -B_z y, \; A_y = 0 \tag{2.80}$$

Eq. (2.79) may be rewritten as:

$$\left[E_s + \frac{(p_x + eB_z y)^2}{2m} + \frac{p_y^2}{2m} + \mathcal{V}(y) \right] \psi(x,y) = E\psi(x,y), \; p_j = -i\hbar \frac{\partial}{\partial j} \tag{2.81}$$

The wavefunctions satisfying Eq. (2.81) read in term of plane-waves:

$$\psi(x,y) = \frac{1}{\sqrt{L}} e^{ikx} \chi(y) \tag{2.82}$$

L denotes the length of the conductor over which the wavefunctions are to be normalized and the transverse function $\chi(y)$ in turn has to satisfy the equation:

$$\left[E_s + \frac{(\hbar k + eB_z y)^2}{2m} + \frac{p_y^2}{2m} + \mathcal{V}(y) \right] \chi(y) = E\chi(y) \tag{2.83}$$

Assuming a parabolic potential $\mathcal{V}(y) = 1/2 \cdot m\omega_0^2 y^2$ Eq. (2.83) may be rewritten as:

$$\left(E_s + \frac{p_y^2}{2m} + \frac{1}{2} m \frac{\omega_0^2 \omega_c^2}{\omega_{c0}^2} y_k^2 + \frac{1}{2} m \omega_{c0}^2 \left[y + \frac{\omega_c^2}{\omega_{c0}^2} y_k \right]^2 \right) \chi(y) = E\chi(y) \tag{2.84}$$

$$\omega_{c0}^2 = \omega_c^2 + \omega_0^2, \; \omega_c = \frac{|e|B_z}{m}, \; y_k = \frac{\hbar k}{eB_z} \tag{2.85}$$

Eq. (2.84) basically represents a one-dimensional Schrödinger equation. Its eigenvalues and eigenfunctions are given by

$$E(n,k) = E_s + \left(n + \frac{1}{2}\right)\hbar\omega_{c0} + \frac{\hbar^2 k^2}{2m}\frac{\omega_0^2}{\omega_{c0}^2} \tag{2.86}$$

$$\chi_{n,k}(y) = u_n\left[q + \frac{\omega_c^2}{\omega_{c0}^2}q_k\right] \tag{2.87}$$

$$q = y\cdot\sqrt{\frac{m\omega_{c0}}{\hbar}},\quad q_k = y_k\cdot\sqrt{\frac{m\omega_{c0}}{\hbar}} \tag{2.88}$$

$$u_n(q) = H_n(q)e^{-\frac{q^2}{2}} \tag{2.89}$$

with $H_n(q)$ being the n-th Hermite polynomial, ω_0 the confinement factor and ω_c the cyclotron frequency. States with different n are said to belong to different *subbands*, with different wavefunctions along the y-axis. These subbands are also known as *transverse modes* in analogy to the modes of an electromagnetic waveguide. The tighter the confinement, the larger ω_0 and thus the mode spacing $\hbar\omega_{c0}$.

Naturally, the assumption of a two-dimensional conductor also involves a confinement along the z-direction which gives rise to the analogous type of subbands along the conductors normal direction. However, this confinement is expected to be very strong, at least in case of surface-supported nanowires. Hence it is assumed that only a single z-subband is occupied. For this reason E_c was replaced by $E_s = E_c + \epsilon_1$ in Eq. (2.79), with ϵ_1 denoting the eigenenergy of the first z-subband.

On first glance the magnetic field simply seems to increase the effective mass by $m \to m \cdot (1 + \omega_c^2/\omega_0^2)$ and thus flattens the dispersion relation. However, by looking at the spatial localization of the wavefunctions it becomes apparent that states carrying current along $+x$ shift to one side of the conductor and states carrying current along $-x$ to the other side. While it also seems reasonable from a classical point of view (Hall effect), it does reduce the spatial overlap between forward and backward propagating states and can thus suppress backscattering due to imperfections. For zero magnetic field the purely electronic subbands with effective mass m are retained.

This treatment now allows for two important conclusions:

- In case the conductance were to be measured purely through this type of conductor, in a contactless way without any boundary layers involved in the entire experiment,[5] the conductance would be exactly equal to the number of these

---

[5] In fact this is possible in principle by IR spectroscopy. The group of Prof. N. Esser at the TU Berlin is currently working on such a measurement.

2.2. Transport theory

modes [114]. This will be further elaborated in the section regarding the Landauer formalism. For the present work the extremely simple possibility of calculating the conductance by counting the number of modes allows to conveniently check the correctness of the yet to be described Green's function approach for simple model cases. While the approach is formally exact this is an important test due to a multitude of numerical conditions that have to be met.

- The above Schrödinger equation can also be solved numerically with the potential $\mathcal{V}(r)$ obtained from DFT as the total effective single-particle potential. This approach will be used in the present work to corroborate some of the considerations regarding the transport properties of nanowires containing impurities.

2.2.1b Degenerate and non-degenerate conductors

At equilibrium conditions the conductor's states are filled up according to the Fermi distribution [106]

$$f_0(E) = (1 + e^{\frac{E-E_F}{k_B T}})^{-1} \tag{2.90}$$

with E_F representing the Fermi energy. There are two limits in which the Fermi distribution inside the band $(E - E_s)$ may be simplified [106]:

- *High temperature or non-degenerate limit* ($e^{\frac{E_s - E_F}{k_B T}} \gg 1$)

$$f_0(E) \approx e^{-\frac{E-E_F}{k_B T}} \tag{2.91}$$

- *Low temperature or degenerate limit* ($e^{\frac{E_s - E_F}{k_B T}} \ll 1$)

$$f_0(E) \approx \vartheta(E_F - E) \tag{2.92}$$

In the present work mainly the low temperature (LT) limit is involved. Then the equilibrium electron density n_s per unit area can be expressed as:

$$n_s = \int dE\, N(E) \cdot f_0(E) \approx \frac{m}{\pi \hbar^2}(E_F - E_s) \tag{2.93}$$

In the LT limit the conductance is determined entirely by Fermi wavevector electrons

$$E_F - E_s = \frac{\hbar^2 k_F^2}{2m} \Rightarrow k_F = \frac{1}{\hbar}\sqrt{2m(E_F - E_s)} = \sqrt{2\pi n_s} \tag{2.94}$$

with the corresponding Fermi velocity $v_f = \hbar k_F m^{-1} = \hbar m^{-1}\sqrt{2\pi n_s}$.

2.2.1c Characteristic length scales

There are three characteristic length scales that determine whether a conductor exhibits classical ohmic behaviour or has to be described using the concepts introduced in the present section. Non-ohmic behaviour emerges if any of these length scales fall in the order of magnitude of the conductor's dimensions [106].

De Broglie wavelength (λ): As a consequence of Eq. (2.94) the Fermi wavenumber increases with $\sqrt{n_s}$. Thus the inverse applies to the corresponding wavelength:

$$\lambda_F = \frac{2\pi}{k_F} = \sqrt{\frac{2\pi}{n_s}} \qquad (2.95)$$

Since in the low temperature limit the current is mostly carried by electrons close to the Fermi energy, this is usually the relevant length scale. I. e. for $n_s = 5 \cdot 10^{11} cm^{-2}$ the Fermi wavelength is $\lambda_F = 35$ nm.

Mean free path (L_m): A single electron moves within an ideal crystal as if in vacuum but with a different mass, the effective mass m. Any deviations from the ideal state, caused, i. e., by impurities, phonons or other electrons, lead to scattering processes that change the electrons' momentum. The momentum relaxation time τ_m is thus related to the collision time τ_c by

$$\tau_m^{-1} \propto \tau_c^{-1} \cdot \alpha_m \qquad (2.96)$$

with α_m denoting the "effectiveness" of an individual collision in changing an electron's momentum. The mean free path L_m that an electron travels until its initial momentum is changed can hence be defined as:

$$L_m = v_F \cdot \tau_m \qquad (2.97)$$

Assuming, i. e., $n_s = 5 \cdot 10^{11}$ cm^{-2} and $\tau_m = 100$ ps one obtains a mean free path of $L_m = 30$ µm.

Phase-relaxation length (L_φ): The concept of phase may be clarified by considering an electron split-beam interference experiment inside a magnetic field [113]. By adjusting the magnetic field's strength one can change the electron's relative phase and thus tune the interference through alternate constructive and destructive cycles. In

2.2. Transport theory

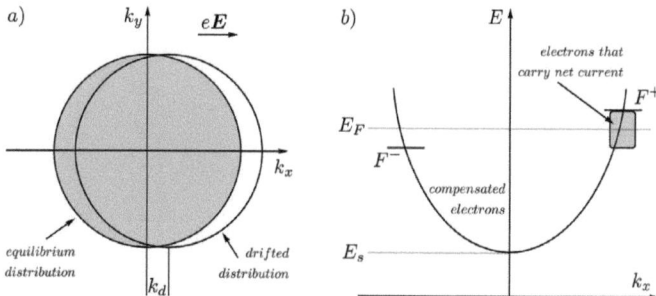

Figure 2.4: *a) In equilibrium at $T \to 0$ K all states within the Fermi circle of radius k_F are occupied. Application of an electric field shifts the Fermi circle along eE. b) States carrying current along $+x$ and $-x$ are filled up to different quasi-Fermi levels F^+ and F^-, respectively. Net current flow only occurs in the interval $[F^-, F^+]$.*

analogy to τ_m a similar relation applies to the phase relaxation time τ_φ

$$\tau_\varphi^{-1} \propto \tau_c^{-1} \cdot \alpha_\varphi \tag{2.98}$$

where α_φ denotes the "effectiveness" of an individual collision in changing an electron's phase. Phase changing scattering processes include dynamic scattering at impurities with internal degrees of freedom, electron-electron and electron-phonon interaction.[6] The phase relaxation time may be considerably larger than the momentum relaxation time, i. e. in the split-beam experiment the two paths may be many times L_m, so that after τ_φ the velocity vector is completely randomized. It is therefore best to express L_φ in terms of the diffusion constant D:

$$L_\varphi = \sqrt{D\tau_\varphi}, \text{ with } D = \frac{1}{2}v_F^2 \tau_m \tag{2.99}$$

2.2.1d Conduction as dynamics of Fermi-energy electrons

The definition of the current density $J = en_s v_d$ as the product of the electron density n_s and drift velocity v_d conveys the impression, that all the conduction electrons drift along and thus contribute to the current. However, in fact for a degenerate conductor

---

[6] Static scattering processes, such as scattering at rigid impurities without internal degrees of freedom, do not contribute to phase relaxation. Considering, i. e., the above split-beam experiment a systematic phase-relationship between the two paths would still exist. This is also in agreement with a slightly more philosophical argument by Feynman [108]: The interference and thus the phase-relationship is destroyed once the observer is able to tell the actual path an individual electron took. The observer would be able to do so if by interacting the electron changed the state of a dynamic scatterer.

at low temperatures ($k_B T \ll E_F - E_s$) the current is non-zero only within a few $k_B T$ around the Fermi energy. Consider a distribution function $f(k)$ that returns the probability that a state at k is occupied. In equilibrium with no electric field the distribution yields [106]

$$f(k) = 1 \; \forall \; \|k\| < k_F \tag{2.100}$$

for all states inside a circle with radius k_F. Application of an electric field causes the entire distribution to shift according to:

$$[f(k)]_{E\neq 0} = [f(k - k_d)]_{E=0}, \quad \frac{\hbar k_d}{m} = v_d = \frac{eE\tau_m}{m} \rightarrow k_d = \frac{eE\tau_m}{\hbar} \tag{2.101}$$

Deep inside the Fermi sea with $\|k\| \ll \|k_F\|$ nothing happens, if $\|k_d\| \ll \|k_F\|$. Only near $+k_F$ states that were empty become filled and around $-k_F$ states that were filled become empty (see Fig. 2.4a). Thus from collective point of view the electric field only moves some electrons from $-k_F$ to $+k_F$. Rewriting the current density as

$$J = e \left[n_s \frac{\|v_d\|}{\|v_F\|} \right] v_F \tag{2.102}$$

implies the current is carried by a small fraction of the total electrons only that moves with the Fermi velocity. An approximate way to visualize the shift of the distribution $f(k)$ is the introduction of quasi-Fermi levels F^+ and F^- for electrons moving parallel and antiparallel to eE, respectively (see Fig. 2.4b). Net current is carried only by electrons in states in the energy range $[F^-, F^+]$. F^+ and F^- can be estimated by

$$F^+ \sim \frac{\hbar^2 (k_F + k_d)^2}{2m}, \quad F^- \sim \frac{\hbar^2 (k_F - k_d)^2}{2m} \tag{2.103}$$

with k_d given by Eq. (2.101). For $\|k_F\| \gg \|k_d\|$ it follows

$$F^+ - F^- \sim \frac{2\hbar \, k_F \cdot k_d}{m} = 2e\tau_m \, E \cdot v_F = 2e \, E \cdot L_m, \tag{2.104}$$

which implies that the separation of the quasi-Fermi levels is proportional to the energy an electron gains in the applied electric field within the mean free path.

2.2.2 Landauer conductance formalism

Consider a piece of conducting material, that is connected across two large contact pads. If the dimensions of this conductor are large compared to the characteristic lengths described in subsection 2.2.1c its conductance G is given by

$$G = \sigma \frac{W}{L} \tag{2.105}$$

2.2. Transport theory

with W and L representing the width and length of the conductor, respectively. The specific conductivity σ is a material parameter that is independent of the conductor's dimensions. From the ohmic point of view the conductance would grow indefinitely with $G \to \infty$ for $L \to 0$. However, experimentally the measured conductance approaches a limit $G \to G_C$ when $L \ll L_m$ becomes significantly shorter than the mean free path. This resistance G_C^{-1} arises at the interfaces between conductor and contact pads. Within the contacts the current is carried by infinitely many transverse modes but inside the conductor only by a handful of modes. The hence required current redistribution gives rise to the observed *contact resistance* G_C^{-1} [106].

2.2.2a Calculation of the contact resistance

G_C may be obtained by calculating the current through a ballistic conductor (that is, a conductor without scattering) for a given bias $\mu_1 - \mu_2$. The contacts in this setup are assumed to be *reflectionless*, meaning that the electrons can exit from the narrow conductor into the wide contacts with a negligible probability of reflection. This is a reasonable assumption as long as the energy is not too close to the bottom of the band [117]. However, the other way around insertion from contact into conductor may exhibit quite large reflections.

This assumption represents a considerable simplification [106]: $k+$ and $k-$ states are occupied only by electrons originating in the left and right contacts, respectively. This is because electrons originating in the left contact populate the $k+$ states and empty reflectionless into the right contact, while electrons originating from the right contact populate the $k-$ states and empty reflectionless into the left contact. *As a consequence there is no causal relationship between the right contact and the $k+$ states nor between the left contact and the $k-$ states.* Hence the quasi-Fermi level F^+ for the $k+$ states is always equal to μ_1 and F^- for the $k-$ states is always equal to μ_2, even when a bias voltage is applied. Thus at $T \to 0K$ the current is exactly equal to that carried by all $k+$ states within $[\mu_1, \mu_2]$.

The states in the conductor belong to different transverse modes, each exhibiting a dispersion relation $E(N,k)$ (cf. Fig. 2.4b) with a cutoff energy $\epsilon_N = E(N, k = 0)$ below which it cannot propagate. The number of transverse modes at an energy E is given by:

$$M(E) = \sum_N \vartheta(E - \epsilon_N), \quad \vartheta(E) = \begin{cases} 1, & E \geq 0 \\ 0, & E < 0 \end{cases} \tag{2.106}$$

The current carried by each transverse mode N can be evaluated separately and then summed up. A single transverse mode in a homogeneous electron gas with n electrons per unit length moving with v, whose $k+$ states are occupied according to a distribution $f^+(E)$, carries a current equal to env. Since the density n associated with a single k state in a conductor of length L amounts to L^{-1}, the current I^+ carried by the $k+$ states is:

$$I^+ = e \frac{1}{L} \sum_k v f^+(E) = e \frac{1}{L} \sum_k \frac{1}{\hbar} \frac{\partial E}{\partial k} f^+(E) \tag{2.107}$$

By assuming periodic boundary conditions and converting the sum to an integral while accounting for spin according to

$$\sum_k \to 2 \cdot \frac{L}{2\pi} \int_k \tag{2.108}$$

the expression for the current I^+ assumes the form

$$I^+ = \frac{2e}{h} \int_\epsilon^\infty dE \, f^+(E) \tag{2.109}$$

Extending this result to multi-mode conductors yields:

$$I^+ = \frac{2e}{h} \int_{-\infty}^{+\infty} dE \, f^+(E) M(E) \tag{2.110}$$

This result is independent of the actual dispersion relation $E(N,k)$. Thus the current carried per mode per unit energy by an occupied state amounts to $-2e/h \approx$ 80nA/meV. Assuming the number of modes $M(E)$ to be constant for $\mu_1 < E < \mu_2$ yields:

$$I = \frac{2e^2}{h} M \frac{\mu_1 - \mu_2}{e} \Rightarrow G_C = \frac{2e^2}{h} M, \; G_C^{-1} \approx \frac{1}{M} \cdot 12.9 \text{k}\Omega \tag{2.111}$$

Thus the contact resistance decreases inversely with the number of modes. For a single-moded ballistic conductor placed between to reflectionless contacts the contact resistance amounts to $\approx$ 12.9 kΩ.

2.2.2b The Landauer formula

In essence in the microscopic regime the Ohmic scaling law $G = \sigma W/L$ is no longer valid. Instead there are two corrections to Ohm's law:

- A *contact resistance* G_C^{-1} that is independent of the conductor's length L
- A *step-like* progression of the conductance with the width W, depending on the *number of transverse modes* $M(E)$

2.2. Transport theory

Both features are incorporated in the Landauer conductance formula [106]:

$$G = \frac{2e^2}{h} M(E) T(E) \tag{2.112}$$

It approaches the calculation of the conductance from a probabilistic point of view. $T(E)$ denotes the average probability that an electron of energy E injected at one end of the conductor will transmit to the other end. If $T(E) = 1$ the correct expression for a ballistic conductor including contact resistance is recovered.

Consider the setup illustrated in Fig. 2.5a). A conductor is connected by two leads to two large contact pads. The leads are assumed to be ballistic conductors, each with $M(E)$ transverse modes. $T(E)$ denotes the average probability that an electron injected into lead 1 (L_1) will eject into lead 2 (L_2). The contacts are again assumed to be reflectionless. As a consequence k_x+ states in L_1 are occupied only by electrons originating from the left contact (C_1) with the potential μ_1. In the same way k_x- states in L_2 are occupied only by electrons originating from the right contact (C_2) with the potential μ_2. For $T \to 0K$ the electron influx from L_1 is given by:

$$I_1^+ = \frac{2e}{h} M(E) [\mu_1 - \mu_2] \tag{2.113}$$

The outflux I_2^+ from L_2 is $T(E) \cdot I_1^+$ with the rest of the flux I_1^- reflected back to C_1:

$$I_2^+ = \frac{2e}{h} M(E) T(E) [\mu_1 - \mu_2], \quad I_1^- = \frac{2e}{h} M(E) (1 - T(E)) [\mu_1 - \mu_2] \tag{2.114}$$

Thus the net current flow I amounts to:

$$I = I_1^+ - I_1^- = I_2^+ = \frac{2e}{h} M(E) T(E) [\mu_1 - \mu_2] \tag{2.115}$$

With the conductance given by

$$\boxed{G = \frac{I}{-\frac{\mu_1 - \mu_2}{e}} = \frac{2e^2}{h} M(E) T(E)} \tag{2.116}$$

the expression for the Landauer conductance (2.112) is obtained. However, so far only the conductance between L_1 and L_2 was evaluated instead of between C_1 and C_2. Fortunately, the assumption of reflectionless contacts determines that calculating the conductance between C_1 and C_2 yields exactly the same result [106].

In any conductor the quasi-Fermi levels F^+ and F^- must be at least slightly different for a net current to flow. However, a contact has ideally an almost infinite number of

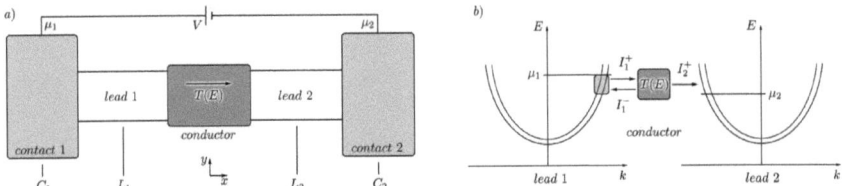

Figure 2.5: *a) Conductor setup for the derivation of the Landauer formula. b) Dispersion relation along k_x with forward and backward traveling currents indicated, respectively.*

modes with an infinitesimal amount of current per mode. Thus F^+ may be assumed to be almost equal to F^-. Naturally, the same does not apply to the leads because they have very few modes and hence cannot be in local equilibrium with the potential μ. For this reason the common argument is that the conductance would have to be evaluated between C_1 and C_2. Fortunately, the derivation of the Landauer conductance does only depend on the energy distribution of the *incoming* electrons at the lead. The energy distribution of any outgoing electrons does not enter the treatment. Since the contacts are reflectionless an electron originating from C_2 can never occupy a k_x+ state, because it originally occupies a k_x- state and ejects reflectionless into L_1. The same applies the other way around. Hence due to the reflectionless contacts the incoming states in each lead are always in equilibrium with the corresponding contact and thus no error is made by evaluating the conductance between L_1 and L_2.

2.2.2c Linear response for non-zero temperatures

The preceding derivation of the Landauer formula applies only for zero temperature, however. This implies that the current is only carried by a single energy channel around the Fermi energy. In general for non-zero temperature transport occurs through many energy channels in the range

$$\mu_1 + n \cdot k_B T > E > \mu_2 - n \cdot k_B T \tag{2.117}$$

while each channel may exhibit a different transmission coefficient $\bar{T}(E) = M(E) \cdot T(E)$. To derive an expression for the current it is now necessary to include electron injection from both contacts. The electron influx per unit energy from L_1 and L_2 is given by

$$i_1^+(E) = \frac{2e}{h} M_1(E) f_1(E) \tag{2.118}$$

$$i_2^-(E) = \frac{2e}{h} M_2(E) f_2(E) \tag{2.119}$$

2.2. Transport theory

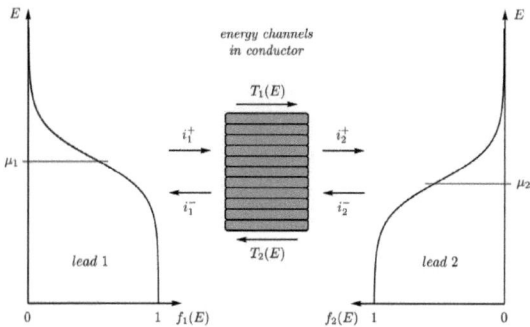

Figure 2.6: *Energy distribution of the incident electrons at non-zero temperature.*

where $f_{1,2}(E)$ denotes the energy distribution of the electrons. The electron outflux per unit energy from L_1 and L_2 is given by

$$i_1^-(E) = (1 - T_1(E))i_1^+(E) + T_2(E)i_2^-(E) \qquad (2.120)$$
$$i_2^+(E) = T_1(E)i_1^+(E) + (1 - T_2(E))i_2^-(E) \qquad (2.121)$$

where $T_{1,2}$ denotes the transmission probabilities from L_1 to L_2 and vice versa, respectively (see Fig. 2.6). Hence the net current $i(E)$ can be written as:

$$i(E) = i_1^+ - i_1^- = i_2^+ - i_2^- \qquad (2.122)$$
$$= T_1 i_1^+ - T_2 i_2^- \qquad (2.123)$$
$$= \frac{2e}{h}[M_1(E)T_1(E)f_1(E) - M_2(E)T_2(E)f_2(E)] \qquad (2.124)$$
$$= \frac{2e}{h}[\bar{T}_1(E)f_1(E) - \bar{T}_2(E)f_2(E)] \qquad (2.125)$$

Under the assumption that no inelastic scattering from one energy to another occurs $\bar{T}_1(E) = \bar{T}_2(E)$. Then the total current amounts to:

$$I = \int dE\, i(E), \quad i(E) = \frac{2e}{h}\bar{T}(E)[f_1(E) - f_2(E)] \qquad (2.126)$$

At $\mu_1 = \mu_2 \Rightarrow f_1(E) = f_2(E)$ the current is zero. For small deviations from this equilibrium Eq. (2.126) implies:

$$\delta I = \frac{2e}{h}\int dE\, \left([\bar{T}(E)]_{eq}\delta[f_1 - f_2] + [f_1 - f_2]_{eq}\delta[\bar{T}(E)]\right) \qquad (2.127)$$

The second term is zero and the first one may be expanded into a Taylor series acording to

$$\delta[f_1 - f_2] \approx [\mu_1 - \mu_2] \left(\frac{\partial f}{\partial \mu}\right)_{eq} = \left(\frac{\partial f_0}{\partial E}\right)[\mu_1 - \mu_2] \quad (2.128)$$

with f_0 representing the Fermi distribution:

$$f_0(E) = \left[\frac{1}{e^{\frac{E-\mu}{k_B T}} + 1}\right]_{\mu = E_F} \quad (2.129)$$

Thus the non-zero temperature conductance in linear response amounts to:

$$\boxed{G = \frac{\delta I}{\frac{\mu_1 - \mu_2}{e}} = \frac{2e^2}{h} \int dE \, \bar{T}(E) \left(-\frac{\partial f_0}{\partial E}\right)} \quad (2.130)$$

In the limit of zero temperature Eq. (2.130) is reduced to the original Landauer conductance expression (2.112).

2.2.3 Landauer conductance from Green's functions

As demonstrated in the previous section the current can be expressed in terms of the transmission function. The transmission function itself may be obtained from the conductor's *scattering matrix* S. This S-matrix allows to characterize a coherent conductor by relating the outgoing wave amplitudes b_i to the incoming wave amplitudes a_i at the different leads according to $b = Sa$. The total number of modes

$$M_T(E) = \sum_i M_i(E) \quad (2.131)$$

is given by the sum over the number of propagating modes $M_p(E)$ at the lead i. Then the S-matrix has the dimension $M_T \times M_T$. Once the S-matrix is known the transmission probability from mode n to m across the conductor is given by the squared magnitude of the corresponding S-matrix element:

$$T_{nm} = |s_{nm}|^2 \quad (2.132)$$

The total transmission function $\bar{T}_{12}(E)$ from lead 1 to lead 2 is obtained by summing over all modes n in lead 1 and all modes m in lead 2:

$$\bar{T}_{12} = \sum_n \sum_m T_{nm} \quad (2.133)$$

It is important to be noted that the current associated with a scattered wave is pro-

2.2. Transport theory

portional to the square of the wavefunction multiplied by the velocity. So while it is customary to define the $\mathcal{S}$-matrix in terms of the current amplitude one can also define a matrix $\mathcal{S}'$ in terms of the wave amplitudes:

$$s'_{nm} = s_{nm}\sqrt{\frac{v_n}{v_m}} \tag{2.134}$$

However, in practical calculations computing the $\mathcal{S}$-matrix directly is usually not the most convenient approach. A more powerful concept is the Green's function formalism, that describes the response at any point inside or outside the conductor due to the excitation at any other point. The $\mathcal{S}$-matrix and thus the transmission function can then be obtained in a simple manner from the Green's function of the systems by the so-called *Fisher-Lee relation*. This is a more sustainable concept because by means of Green's functions one can also account for electron-electron or electron-phonon interactions, which $\mathcal{S}$-matrices alone cannot perform.

2.2.3a Green's function formalism

Whenever a response r is related to an excitation e by a differential operator $\mathcal{D}$, with

$$\mathcal{D}r = s \tag{2.135}$$

a Green's function $\mathcal{G}$ can be defined and the response expressed according to [105]:

$$r = \mathcal{D}^{-1}s = \mathcal{G}s, \; \mathcal{G} \equiv \mathcal{D}^{-1} \tag{2.136}$$

For a system described by the Hamiltonian $\mathcal{H}$ the problem can also be expressed as

$$[E - \mathcal{H}]\psi = S \tag{2.137}$$

with S representing an excitation term due to the wave incident from one of the leads. Thus the Green's function is given by

$$\mathcal{G} = [E - \mathcal{H}]^{-1} \tag{2.138}$$

with the subband energy E_s incorporated into the potential in the Hamiltonian:

$$\mathcal{H} = \frac{(i\hbar\nabla + e\mathbf{A})^2}{2m} + V(\mathbf{r}) \tag{2.139}$$

Retarded and advanced Green's function: As the inverse of a differential operator the Green's function is only uniquely specified if the boundary conditions are set.

Considering a simple example of a one-dimensional wire with a constant potential V_0 and zero magnetic field $\mathcal{G}$ can be written as [106]:

$$\mathcal{G} = \left[E - V_0 + \frac{\hbar^2}{2m} \frac{\partial^2}{\partial x^2} \right] \quad (2.140)$$

By means of the Green's function the problem can be expressed in an analogous way to the Schrödinger equation (2.142)

$$\left[E - V_0 + \frac{\hbar^2}{2m} \frac{\partial^2}{\partial x^2} \right] \mathcal{G}(x, x') = \delta(x - x') \quad (2.141)$$

$$\left[E - V_0 + \frac{\hbar^2}{2m} \frac{\partial^2}{\partial x^2} \right] \psi(x) = 0 \quad (2.142)$$

with the exception of the $\delta(x - x')$ term. Hence the Green's function $\mathcal{G}(x, x')$ can be visualized as the wavefunction at x resulting from a unit excitation at x'. The solution of Eq. (2.141) assumes the form

$$\mathcal{G}(x, x') = A^+ e^{+ik(x-x')}, \; x > x' \quad (2.143)$$
$$\mathcal{G}(x, x') = A^- e^{-ik(x-x')}, \; x < x' \quad (2.144)$$
$$k = \frac{\sqrt{2m(E - V_0)}}{\hbar} \quad (2.145)$$

with the amplitudes A^+ and A^-. In order to satisfy Eq. (2.141) at $x = x'$ $\mathcal{G}$ must be continous, while its derivative must be discontinous by $2m/\hbar$:

$$[\mathcal{G}(x, x')]_{x=x'^+} = [\mathcal{G}(x, x')]_{x=x'^-} \quad (2.146)$$

$$\left[\frac{\partial \mathcal{G}(x, x')}{\partial x} \right]_{x=x'^+} - \left[\frac{\partial \mathcal{G}(x, x')}{\partial x} \right]_{x=x'^-} = \frac{2m}{\hbar^2} \quad (2.147)$$

As a consequence the amplitudes are given by:

$$A^+ = A^- = -\frac{i}{\hbar v}, \; v = \frac{\hbar k}{m} \quad (2.148)$$

However, it is important to note that there are two possible solutions that satisfy Eq. (2.141):

$$\mathcal{G}^r(x, x') = -\frac{i}{\hbar v} e^{+ik|x-x'|} \quad (2.149)$$

$$\mathcal{G}^a(x, x') = +\frac{i}{\hbar v} e^{-ik|x-x'|} \quad (2.150)$$

2.2. Transport theory

These solutions are called the *retarded* Green's function $\mathcal{G}^r$ and the *advanced* Green's function $\mathcal{G}^a$, which correspond to different boundary conditions: $\mathcal{G}^r$ describes outgoing waves originating at x', while $\mathcal{G}^a$ represents incoming waves disappearing at x'. Usually these boundary conditions are incorporated into the equation itself by introducing an infinitesimal imaginary part $i\eta$ into the energy:

$$\mathcal{G}^r = [E - \mathcal{H} + i\eta]^{-1}, \quad \eta \to 0^+ \qquad (2.151)$$
$$\mathcal{G}^a = [E - \mathcal{H} - i\eta]^{-1}, \quad \eta \to 0^+ \qquad (2.152)$$

In case of $\mathcal{G}^r$ the term $i\eta$ introduces a small positive imaginary component to the wavenumber k, causing the advanced function to grow indefinitely. Since a proper solution must be bounded the retarded function is the only acceptable solution. In analogy the only physically acceptable solution for $\mathcal{G}^a$ is the advanced function.

Extension to the multi-moded case: The Green's function $\mathcal{G}(x,y;x',y')$ represents the wavefunction at (x,y) due to an excitation at $x = x'$, $y = y'$. It can be written in the form [106]

$$\mathcal{G}^r(x,y;x',y') = \sum_m A_m^\pm \chi_m(y) e^{ik_m|x-x'|} \qquad (2.153)$$

with the transverse mode wavefunctions $\chi_m(y)$ satisfying the equation

$$\left[-\frac{\hbar^2}{2m} \frac{\partial^2}{\partial y^2} + V(y) \right] \chi_m(y) = \epsilon_{m,0} \chi_m(y) \qquad (2.154)$$

Since the $\chi_n(y)$ satisfy the same equation with different eigenvalues they are orthogonal:

$$\int dy\, \chi_m(y) \chi_n(y) = \delta_{mn} \qquad (2.155)$$

In analogy to the previous procedure the amplitudes are given by

$$A_m^+ = A_m^- = -\frac{i}{\hbar v_m} \chi_m(y') \qquad (2.156)$$

and thus the Green's function may be written as:

$$\mathcal{G}^r(x,y;x',y') = \sum_m -\frac{i}{\hbar v_m} \chi_m(y) \chi_m(y') e^{ik_m|x-x'|} \qquad (2.157)$$

$$k_m = \frac{\sqrt{2m(E - \epsilon_{m,0})}}{\hbar}, \quad v_m = \frac{\hbar k_m}{m} \qquad (2.158)$$

2.2.3b Fisher-Lee relation

The knowledge of the Green's function allows to calculate the S-matrix elements in a straightforward manner by means of the Fisher-Lee relation [118]. Consider a conductor connected to a set of leads, with the interface between lead 1 and lead 2, and with the conductor defined by $x_{1,2} = 0$, respectively. $\mathcal{G}_{12}^r$ denotes the Green's function between points located on the planes $x_1 = 0$ and $x_2 = 0$ [106]:

$$\mathcal{G}_{12}^r(y_1, y_2) \equiv \mathcal{G}^r(x_1 = 0, y_1; x_2 = 0, y_2) \qquad (2.159)$$

The unit excitation at $x_1 = 0$ gives rise to a wave of amplitude A_1^- from and A_1^+ toward the conductor, with the latter being scattered by the conductor. For the moment neglecting the transverse dimension of the leads $\mathcal{G}_{12}^r$ can thus be written as:

$$\mathcal{G}_{12}^r = \delta_{12} A_1^- + s_{12}' A_1^+ \qquad (2.160)$$

With the help of Eq. (2.134) and (2.148)

$$A_1^+ = A_1^- = -\frac{i}{\hbar v_p}, \quad s_{12}' = s_{12}\sqrt{\frac{v_1}{v_2}} \qquad (2.161)$$

the S-matrix elements are obtained from the Green's function $\mathcal{G}_{12}^r$ according to:

$$s_{12} = -\delta_{12} - i\hbar\sqrt{v_1 v_2}\,\mathcal{G}_{12}^r \qquad (2.162)$$

Extension to multi-moded leads: Instead of Eq. (2.160) the Green's function $\mathcal{G}_{12}^r$ can be expressed as:

$$\mathcal{G}_{12}^r(y_1, y_2) = \sum_m \sum_n [\delta_{nm} A_m^- + s_{nm}' A_m^+] \chi_n(y_2) \qquad (2.163)$$

From Eq. (2.134) and (2.156) follow:

$$\mathcal{G}_{12}^r(y_1, y_2) = \sum_m \sum_n -\frac{i}{\hbar\sqrt{v_n v_m}} \chi_n(y_2)[\delta_{12} + s_{12}] \chi_m(y_1) \qquad (2.164)$$

Multiplying Eq. (2.164) by $\chi_m(y_1)\chi_m(y_2)$ and subsequent integrating over y_1, y_2 while using the orthogonality relation Eq. (2.155) yields:

$$\boxed{s_{mn} = -\delta_{mn} + i\hbar\sqrt{v_m v_n} \int\!\!\int dy_2 dy_1\, \chi_n(y_2)[\mathcal{G}_{12}^r(y_1, y_2)]\chi_m(y_1)} \qquad (2.165)$$

2.2. Transport theory

2.2.3c Tight-binding approach

Calculating the Green's function of an arbitrarily shaped conductor requires solving the differential equation for the Green's function for arbitrary $V(r)$ and $A(r)$:

$$[E - \mathcal{H}(r) + i\eta]\mathcal{G}^r(r,r') = \delta(r-r') \tag{2.166}$$

$$\mathcal{H}(r) = \frac{(i\hbar\nabla + eA(r))^2}{2m} + V(r) \tag{2.167}$$

For numerical evaluation the spatial coordinates are discretized, so that the Green's function turns into a matrix

$$\mathcal{G}^r(r,r') \to \mathcal{G}^r(i,j) \tag{2.168}$$

with i, j denoting points on a discrete lattice with an equidistant spacing a. Hence the differential equation (2.166) turns into a matrix equation

$$[(E + i\eta)\mathcal{I} - \mathcal{H}]\mathcal{G}^r = \mathcal{I} \tag{2.169}$$

where $\mathcal{I}$ represents the identity matrix and $\mathcal{H}$ now denotes the matrix representation of the system's Hamiltonian. In principle by inverting the matrix $[(E + i\eta)\mathcal{I} - \mathcal{H}]$ numerically, one could now obtain the desired Green's function $\mathcal{G}^r(i,j)$. However, throughout the present treatment it was always assumed that the conductor is attached to leads, which allow the electrons to empty *reflectionless* into the contacts. This presents a problem because as a consequence the resulting matrix to be inverted is *infinite-dimensional*. Unfortunately simply truncating the matrix would mean to introduce *finite* leads with *fully reflecting* boundaries [106].

To still allow for a way to perform the necessary truncation and thus enable the numerical calculation of $\mathcal{G}^r(i,j)$ it is highly useful to employ the so-called *tight-binding* scheme. This approach expresses the electronic wavefunctions in terms of localized atomic orbitals that are centered on the atoms constituting the system. Consider a 2-terminal setup consisting of a conductor that is attached to two semi-infinite (and thus reflectionless) leads on either side. Within the tight-binding scheme – and subsequently within a *localized basis set* – the total Green's function of this system can be partitioned into submatrices that correspond to the individual parts of the system (cf. Fig. 2.7):

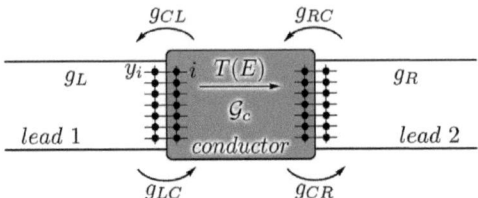

Figure 2.7: *A conductor described by $\mathcal{G}_C$ connected across two semi-infinite leads described by g_L, g_R through coupling matrices g_{LC}, g_{CR}. The points y_i are adjacent to the points i.*

$$\begin{pmatrix} g_L & g_{LC} & g_{LCR} \\ g_{CL} & \mathcal{G}_C & g_{CR} \\ g_{LRC} & g_{RC} & g_R \end{pmatrix} = \begin{pmatrix} E\mathcal{I} - h_L & -h_{LC} & 0 \\ -h_{LC}^\dagger & E\mathcal{I} - \mathcal{H}_C & -h_{CR} \\ 0 & -h_{CR}^\dagger & E\mathcal{I} - h_R \end{pmatrix}^{-1} \quad (2.170)$$

$(E\mathcal{I} - \mathcal{H}_C)$ represents the Green's function of the finite isolated conductor without any coupling to the leads, $(E\mathcal{I} - h_{\{L,R\}})$ denotes the Green's function of the semi-infinite leads and h_{LC}, h_{CR} are the coupling matrices between the conductor and the leads. Capital and lower case letters denote finite- and infinite-dimensional matrices, respectively. In Eq. (2.170) zero coupling is assumed between left and right leads by setting $g_{LRC} = g_{LCR} = 0$.[7] From Eq. (2.170) it is straightforward to obtain an explicit expression for $\mathcal{G}_C$ [106]:

$$\mathcal{G}_C(E) = (E\mathcal{I} - \mathcal{H}_C - \Sigma_L(E) - \Sigma_R(E))^{-1} \quad (2.171)$$

$$\Sigma_L(E) = h_{LC}^\dagger (E\mathcal{I} - h_L)^{-1} h_{LC}, \quad \Sigma_R(E) = h_{RC}(E\mathcal{I} - h_R)^{-1} h_{RC}^\dagger \quad (2.172)$$

The terms $\Sigma_L(E)$, $\Sigma_L(E)$ can be viewed as effective Hamiltonians arising from the interaction between the conductor and the leads. A similar term is widely used throughout solid state physics to describe electron-electron and electron-phonon interactions and is called *self-energy*. In analogy these terms are called the self-energies due to the leads. It is to be noted that no approximation is involved in arriving at Eq. (2.171) from Eq. (2.166). The treatment is still formally exact. To employ Eq. (2.171) the Green's functions $g^r_{\{L,R\}} = (E\mathcal{I} - h_{\{L,R\}})^{-1}$ for the isolated leads have to be calculated. For a semi-infinite wire $g^r_{\{L,R\}}$ on a discrete lattice is given by

$$g^r_{\{L,R\}}(y_i, y_j) = -\frac{1}{t} \sum_m \chi_m^{\{L,R\}}(y_i) e^{ik_m a} \chi_m^{\{L,R\}}(y_j), \quad t \equiv \frac{\hbar^2}{2ma^2} \quad (2.173)$$

with a denoting the spacing of the discrete lattice. Inserting Eq. (2.173) into Eq. (2.172)

---

[7] Hence zero coupling also has to be ensured in practical calculations and represents one among several important convergence parameters.

2.2. Transport theory

yields the desired expression for the self-energy:

$$\Sigma^{r}_{\{L,R\}}(y_i, y_j) = -t \sum_{m} \chi^{\{L,R\}}_{m}(y_i) e^{ik_m a} \chi^{\{L,R\}}_{m}(y_j) \qquad (2.174)$$

By inserting Eq. (2.174) into Eq. (2.171) and applying the Fisher-Lee relation (2.165) the transmission function can be obtained in a very compact form:

$$\boxed{T(E) = \text{Tr}[\Gamma_L \mathcal{G}^{r}_{C} \Gamma_R \mathcal{G}^{a}_{C}]} \qquad (2.175)$$

Tr denotes the trace of the matrix and the $\Gamma_{\{L,R\}}$ represent coupling functions that describe the coupling of the conductor to the leads. The elements of the matrices $\Gamma_{\{L,R\}}$ are given by:

$$\Gamma_{\{L,R\}}(i,j) = \sum_{m} \chi^{\{L,R\}}_{m}(y_i) \frac{\hbar v_m}{a} \chi^{\{L,R\}}_{m}(y_j) \qquad (2.176)$$

With Eq. (2.174) and the relation $\hbar v = \partial E/\partial k = 2at \cdot \sin(ka)$ for the discrete lattice the matrices $\Gamma_{\{L,R\}}$ can be rewritten as

$$\boxed{\Gamma_{\{L,R\}} = i \left[\Sigma^{r}_{\{L,R\}}(E) - \Sigma^{a}_{\{L,R\}}(E) \right]} \qquad (2.177)$$

where the advanced self-energy $\Sigma^{a}_{\{L,R\}}$ is the Hermitian conjugate of the retarded self-energy $\Sigma^{r}_{\{L,R\}}$. Ultimately the core of the problem lies in the calculation of the isolated conductor's Green's function $\mathcal{G}_C$ and the self-energies of the semi-infinite leads $\Sigma^{r}_{\{L,R\}}$.

2.2.4 Bridging the gap to DFT

As introduced in section 2.1.5b the present work employs the supercell approach for describing any systems at DFT level. The following sections describe a computational scheme, that allows to calculate the required Green's functions and self-energies from state of the art plane-wave DFT calculations within the supercell approach. *Wannier functions* will be used as a means to connect the Green's function tight-binding approach to plane-wave DFT electronic structure calculations. Both the Bloch functions within DFT and the Wannier functions span exactly the same function space, in this case a Hilbert space. Hence this scheme allows a very efficient description of electron transport within the Landauer formalism at full *first principles* DFT precision, simultaneously.

2.2.4a Transmission within the supercell approach

Any solid or surface can be represented as an infinite or semi-infinite stack of principal layers with nearest-neighbour interactions [119, 120], respectively. A left lead – conductor – right lead (LCR) setup (cf. Fig. 2.8) can be considered as one principal layer, that describes the conductor, sandwiched between two semi-infinite stacks of principal layers describing the leads. Within this approach the matrix elements of Eq. (2.169) between layer orbitals yield a series of matrix equations

$$\begin{aligned}(E - \mathcal{H}_{00})\mathcal{G}_{00} &= I + \mathcal{H}_{01}\mathcal{G}_{10} \\ (E - \mathcal{H}_{00})\mathcal{G}_{10} &= I + \mathcal{H}_{01}^{\dagger}\mathcal{G}_{00} + \mathcal{H}_{01}\mathcal{G}_{20} \\ &\cdots \\ (E - \mathcal{H}_{00})\mathcal{G}_{n0} &= I + \mathcal{H}_{01}^{\dagger}\mathcal{G}_{n-1,0} + \mathcal{H}_{01}\mathcal{G}_{n+1,0}\end{aligned} \quad (2.178)$$

with the finite dimension matrices $\mathcal{H}_{nm}$, $\mathcal{G}_{nm}$ given by the matrix elements between the layer orbitals. Assuming a bulk system for the moment, with $\mathcal{H}_{00} = \mathcal{H}_{11} = ...$, $\mathcal{H}_{01} = \mathcal{H}_{12} = ...$, the series can be reduced by expressing the Green's function of an individual layer in terms of the Green's function of the preceding or following layer [121, 122]. By introducing the transfer matrices $\mathcal{T}$, $\tilde{\mathcal{T}}$, with $\mathcal{G}_{10} = \mathcal{T}\mathcal{G}_{00}$, $\mathcal{G}_{00} = \tilde{\mathcal{T}}\mathcal{G}_{10}$, the bulk Green's function can be expressed as [123]:

$$\mathcal{G}(E) = (E - \mathcal{H}_{00} - \mathcal{H}_{01}\mathcal{T} - \mathcal{H}_{01}^{\dagger}\tilde{\mathcal{T}})^{-1} \quad (2.179)$$

The required transfer matrices can be obtained from the Hamiltonian matrix elements by a simple recursive algorithm that usually requires no more than 6 iterations until it converges [119, 120]:

$$\begin{aligned} \mathcal{T} &= t_0 + \tilde{t}_0 t_1 + \tilde{t}_0 \tilde{t}_1 t_2 + ... + \tilde{t}_0 \tilde{t}_1 \tilde{t}_2 ... t_n \\ \tilde{\mathcal{T}} &= \tilde{t}_0 + t_0 \tilde{t}_1 + t_0 t_1 \tilde{t}_2 + ... + t_0 t_1 t_2 ... \tilde{t}_n \\ t_i &= (I - t_{i-1}\tilde{t}_{i-1} - \tilde{t}_{i-1}t_{i-1})^{-1} t_{i-1}^2 \\ \tilde{t}_i &= (I - t_{i-1}\tilde{t}_{i-1} - \tilde{t}_{i-1}t_{i-1})^{-1} \tilde{t}_{i-1}^2 \\ t_0 &= (E - H_{00})^{-1} H_{01}^{\dagger} \\ \tilde{t}_0 &= (E - H_{00})^{-1} H_{01} \end{aligned} \quad (2.180)$$

From the comparison of Eq. (2.179) and (2.171) the conductor's Hamiltonian and the self energies can be obtained as:

$$\boxed{\mathcal{H}_C = \mathcal{H}_{00},\ \Sigma_L = \mathcal{H}_{01}^{\dagger}\tilde{\mathcal{T}},\ \Sigma_R = \mathcal{H}_{01}\mathcal{T}} \quad (2.181)$$

The coupling functions can be obtained from the the transfer matrices and Hamilto-

2.2. Transport theory

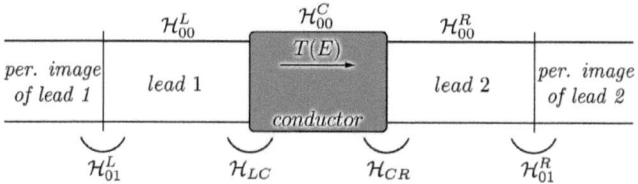

Figure 2.8: *Schematic representation of the left lead – conductor – right lead system within the principal layer approach. Elements of the corresponding Hamiltonian are indicated.*

nian matrix elements as well [125]:

$$\Gamma_L = -\text{Im}(\mathcal{H}_{01}^\dagger \mathcal{T}), \; \Gamma_R = -\text{Im}(\mathcal{H}_{01} \bar{\mathcal{T}}) \qquad (2.182)$$

Transmission through a left lead – conductor – right lead system: From now on the assumption of a bulk system ($\mathcal{H}_{00} = \mathcal{H}_{11} = ...$, $\mathcal{H}_{01} = \mathcal{H}_{12} = ...$) is dropped in order to generalize the description to the experimentally more realistic geometry of a left lead – conductor – right lead (LCR) system (cf. Fig. 2.8). Hence the individual principal layers may now contain different media. As a consequence this case requires also a description of the interface regions in terms of Green's functions. Within the framework of *surface Green's function matching* (SGFM) theory the Green's functions of the interface regions may be calculated from the bulk Green's functions of the isolated systems [123, 124]. By calculating the transmitted and reflected amplitudes of a unit excitation propagating from one medium into the other, it can be shown that the surface Green's function obeys the equation [123]

$$\mathcal{G}_{LCR} = \begin{pmatrix} \mathcal{G}_L & \mathcal{G}_{LC} & \mathcal{G}_{LR} \\ \mathcal{G}_{CL} & \mathcal{G}_C & \mathcal{G}_{CR} \\ \mathcal{G}_{RL} & \mathcal{G}_{RC} & \mathcal{G}_R \end{pmatrix}$$
$$= \begin{pmatrix} E - \mathcal{H}_{00}^L - (\mathcal{H}_{01}^L)^\dagger \mathcal{T} & -\mathcal{H}_{LC} & 0 \\ -\mathcal{H}_{CL} & E - \mathcal{H}_C & -\mathcal{H}_{CR} \\ 0 & -\mathcal{H}_{RC} & E - \mathcal{H}_{00}^R - \mathcal{H}_{01}^R \bar{\mathcal{T}} \end{pmatrix}^{-1} \qquad (2.183)$$

with $\mathcal{H}_{nm}^{\{L,R\}}$ denoting the Hamiltonian's matrix elements between the layer orbitals in the left and right leads, respectively (cf. Fig. 2.8). The transfer matrices can be computed as described previously. $\mathcal{H}_{LC}, \mathcal{H}_{CR}$ represent the coupling matrices between the conductor and the left and right leads, respectively. It is straightforward to obtain from Eq. (2.183) the Green's function $\mathcal{G}_C$ again in the form $\mathcal{G}_C = (E - \mathcal{H}_C - \Sigma_L - \Sigma_R)^{-1}$ with the self-energies given by [125]:

$$\boxed{\begin{aligned}\Sigma_L(E) &= \mathcal{H}_{LC}^\dagger(E - \mathcal{H}_{00}^L - (\mathcal{H}_{01}^L)^\dagger \tilde{T}_L)^{-1}\mathcal{H}_{LC} \\ \Sigma_R(E) &= \mathcal{H}_{CR}(E - \mathcal{H}_{00}^R - \mathcal{H}_{01}^R T_R)^{-1}\mathcal{H}_{CR}^\dagger\end{aligned}} \qquad (2.184)$$

Thus the transmission in the general LCR geometry can be obtained from Eq. (2.171), (2.175), (2.177) and (2.184). The Green's function $\mathcal{G}_C$ also contains direct information regarding the electronic spectrum via the spectral density of states:

$$N(E) = -\frac{1}{\pi}\text{Im}[\text{Tr}(\mathcal{G}_C(E))] \qquad (2.185)$$

Within DFT the principal layers described in this calculational scheme can be modeled by separate supercell calculations for the conductor region and each of the leads, respectively. To ensure a complete *ab initio* description of the interface regions, the conductor supercell should contain a sufficiently large region of the leads as well. Sufficient in this context means that the electronic structure at the boundaries of the conductor supercell is converged up to reaching the bulk electronic structure of the leads.

However, so far the chain of principal layers was always assumed to be strictly one-dimensional (1D). This approach can be extended to a truly three-dimensional (3D) description by effectively describing the three-dimensional case as comprised of an infinite number of non-interacting one-dimensional chains. In terms of practical DFT calculations this means that a sufficiently large number $k_\perp$ of k-points perpendicular to the transport direction has to be taken into account. The total transmittance can then be calculated as the weighted sum over the individual transmittances along the $k_\perp$ parallel paths, according to:

$$T(E) = \sum_{k_\perp} w_{k_\perp} T_{k_\perp}(E) \qquad (2.186)$$

2.2.4b Real-space basis set: Wannier functions

The calculation of the ground-state electronic structure for the individual principal layers can be performed employing the DFT supercell approach, as described in the previous section. However, the calculation of the Green's functions relies on a tight-binding scheme and thus a *localized orbital* basis set. On the other hand plane-wave DFT utilizes Bloch orbitals, which are intrinsically delocalized. The transformation of Bloch functions into *Wannier functions* (WF) represents a natural choice for a localized basis set, because both function types still span exactly the same Hilbert space of

2.2. Transport theory

the Hamiltonian's eigenfunctions. Hence they allow to bridge plane-wave electronic structure and lattice Green's function calculations without any principal loss of accuracy. A Wannier function $w_{n\mathbf{R}}(\mathbf{r})$ at the Bravais lattice vector $\mathbf{R}$ is defined via a unitary transformation of the Bloch functions $\psi_{n\mathbf{k}}(\mathbf{r})$ of the n-th band

$$w_{n\mathbf{R}}(\mathbf{r}) = \frac{V}{(2\pi)^3} \int_{BZ} d\mathbf{k}\, \psi_{n\mathbf{k}}(\mathbf{r}) e^{-i\mathbf{k}\cdot\mathbf{R}} \qquad (2.187)$$

with V denoting the volume of the unit cell and the integration being performed over the entire Brillouin zone. The WFs as defined above form an orthonormal basis set. Any two for a given n and different $\mathbf{R}, \mathbf{R}'$ are simply translational images of each other. However, this definition does not lead to a unique set of WFs due to the $\mathbf{k}$-dependent arbitrary phase factor of the Bloch functions and the invariancy of the electronic structure problem as opposed to the transformation $\psi_{n\mathbf{k}} \to e^{i\phi_n(\mathbf{k})}\psi_{n\mathbf{k}}$. Besides the arbitrary phase factor $\phi_n(\mathbf{k})$ there is also a more fundamental gauge freedom arising from the many-body wavefunction actually being a Slater-determinant. A unitary transformation does not change the manifold and thus leaves the total energy and charge density unchanged as well. Hence starting with a set of N Bloch functions with periodic parts $u_{n\mathbf{k}}$ one can construct an infinite number of sets of N WFs which all exhibit different spatial characteristics:

$$w_{n\mathbf{R}}(\mathbf{r}) = \frac{V}{(2\pi)^3} \int_{BZ} d\mathbf{k}\, \left[\sum_m U_{mn}^{(\mathbf{k})} \psi_{m\mathbf{k}}(\mathbf{r}) \right] e^{-i\mathbf{k}\cdot\mathbf{R}} \qquad (2.188)$$

The gauge freedom of phase is included in the unitary matrices $U^{(\mathbf{k})}$ as well. To utilize these Wannier functions as a basis set for the tight-binding scheme it is necessary to ensure a degree of localization that excludes any but nearest-neighbour interactions between the principal layers, as assumed in the derivation of the Green's function formalism. Hence a computational scheme is required for identifying those $U_{mn}^{(\mathbf{k})}$ that transform the Bloch eigenstates into those WFs with the narrowest spatial distribution.

2.2.4c Localization procedure

Such a scheme that allows to identify those $U_{mn}^{(\mathbf{k})}$ that lead to *maximally localized Wannier functions* was proposed by Marzari and Vanderbilt [126]. A measure of the WF's spatial localization is given by the *spread* operator Ω

$$\Omega = \sum_n [\langle r^2 \rangle_n - \langle \mathbf{r} \rangle_n^2] \qquad (2.189)$$

with the summation perfomed over a selected group of bands and:

$$\langle r \rangle_n = [\langle Rn|r|Rn \rangle]_{R=0} = \langle 0n|r|0n \rangle \qquad (2.190)$$

$$\langle r^2 \rangle_n = [\langle Rn|r^2|Rn \rangle]_{R=0} = \langle 0n|r^2|0n \rangle \qquad (2.191)$$

Since the spread Ω depends on the choice of $U_{mn}^{(k)}$ any arbitrary set of $U_{mn}^{(k)}$ can be evolved until the stationary case is reached:

$$\frac{\partial \Omega_k}{\partial U^{(k)}} = 0 \qquad (2.192)$$

The resulting matrices $U^{(k),ML}$ transform the Bloch functions $\psi_{nk}(r)$ into maximally localized Wannier functions according to Eq. (2.188). Restricting the treatment to the case of k-point mesh calculations Eq. (2.192) can be evaluated employing finite differences in reciprocal space. For this purpose the expectation values $\langle r \rangle, \langle r^2 \rangle$ can be expressed as [130]

$$\langle 0n|r|0n \rangle = i\frac{1}{N}\sum_k e^{ik\cdot R}\langle u_{kn}|\nabla_k|u_{kn}\rangle \qquad (2.193)$$

$$\langle 0n|r|0n \rangle = \frac{1}{N}\sum_k e^{ik\cdot R}\langle u_{kn}|\nabla_k^2|u_{kn}\rangle \qquad (2.194)$$

with $|u_{kn}\rangle = e^{-ik\cdot r}|\psi_{kn}\rangle$ representing the periodic parts of the Bloch functions. The *overlap matrix* between Bloch functions can be defined by

$$M_{mn}^{(k,b)} = \langle u_{mk}|u_{nk+b}\rangle = \langle \psi_{mk}|e^{-ikb}|\psi_{nk+b}\rangle \qquad (2.195)$$

where b denotes vectors that connect a mesh point to its nearest neighbours. These overlap matrices allow to express the expectation values $\langle r \rangle, \langle r^2 \rangle$ in a form that is usable in a practical localization algorithm

$$\langle r \rangle_n = -\frac{1}{N}\sum_{k,b} w_b \, \Im \, \ln M_{nn}^{(k,b)} \qquad (2.196)$$

$$\langle r^2 \rangle_n = \frac{1}{N}\sum_{k,b} w_b \left[\left(1 - |M_{nn}^{(k,b)}|^2\right) + \left(\Im \, \ln M_{nn}^{(k,b)}\right)^2\right] \qquad (2.197)$$

with w_b denoting the weights of the b-vectors, which satisfy the completeness condition $\sum_b w_b b_\alpha b_\beta = \delta_{\alpha\beta}$. Inserting the above expressions into Eq. (2.189) yields an expression for the spread operator in terms of the overlap matrices $M_{mn}^{(k,b)}$. To calculate the gradient in Eq. (2.192) consider the infinitesimal unitary transforma-

2.2. Transport theory

tion $U_{mn}^{(k)} = \delta_{mn} + dW_{mn}^{(k)}$, where dW represents an infinitesimal anti-unitary matrix $dW^\dagger = -dW$. This transformation rotates the wavefunctions according to Eq. (2.188) into $|u_{kn}\rangle \to |u_{kn}\rangle + \sum_m dW_{mn}^{(k)} |u_{km}\rangle$. The following expression can then be obtained, that is straightforward to implement [126]:

$$G^{(k)} = \frac{\partial \Omega}{dW^{(k)}} = 4 \sum_b w_b \left(\frac{R^{(k,b)} - R^{(k,b)\dagger}}{2} - \frac{T^{(k,b)} - T^{(k,b)\dagger}}{2i} \right) \quad (2.198)$$

$$R_{mn}^{(k,b)} = M_{mn}^{(k,b)} M_{nn}^{(k,b)*}, \quad T_{mn}^{(k,b)} = \frac{M_{mn}^{(k,b)}}{M_{nn}^{(k,b)}} \left[\Im \ln M_{nn}^{(k,b)} + \mathbf{b} \cdot \langle \mathbf{r} \rangle_n \right] \quad (2.199)$$

It is to be noted that the entire expression $G^{(k)}$ is a function of the overlap matrices $M_{mn}^{(k,b)}$, which is numerically very convenient. The minimization itself can be conducted by generic steepest descent or conjugate gradients algorithms. Only the overlap and unitary matrices have to be calculated in each step, scaling as $\mathcal{O}(N^3)$. Note that neither the wavefunctions have to be updated nor the actual Wannier functions themselves need to be calculated. The spread of each individual Wannier function represents a very important convergence parameter. Since the Green's function formalism assumes a localized orbital basis set with nearest-neighbour interactions only, it must be ensured that the degree of localization for each Wannier function is sufficient to meet this criterion.

Disentangle-procedure and frozen-states: However, the scheme described so far works only for *isolated groups* of bands. A band is called isolated in this context if it does not become degenerate with any other band anywhere in the Brillouin zone. Generally this is not the case and it is necessary to extend the calculational scheme to the case of mixed or *entangled* bands, i. e. bands exhibiting crossings with other bands. The problem is that any arbitrary choice of states inside a given energy window affects the localization properties of the Wannier functions. Consider, i. e., two bands that are degenerate at some k-point. The DFT band structure alone allows no inference whether the bands simply touch or actually cross. Assuming either case clearly leads to Wannier functions with very different localization properties. This problem is solved by the introduction of an additional *disentanglement*-procedure [127], where the spread Ω is separated into:

$$\Omega = \tilde{\Omega} + \Omega_I \quad (2.200)$$

Only $\tilde{\Omega}$ can be minimized by evolving the $U^{(k)}$. Ω_I has to be minimized by selecting which states are supposed to form a band, or in other words by selecting the optimal subspace $\mathcal{S}(k)$. The procedure begins by choosing a prescribed energy window within the bands are to be disentangled. The energy window has to contain $N_k \geq N$ bands at every k-point, with N denoting the number of desired Wannier functions. This defines the N_k-dimensional Hilbert space $\mathcal{F}(k)$ spanned by the states u_{kn} inside the energy window. For $N_k = N$ there is nothing to do for the disentangle procedure. If $N_k > N$ the optimal subspace $\mathcal{S}(k) \subseteq \mathcal{F}(k)$ needs to be identified, that minimizes Ω_I.

Sometimes it is desirable to treat the states in a small range of interest (usually around the Fermi energy) as frozen to obtain Wannier functions that correspond exactly to these states. This is possible by excluding these states from the disentangle-procedure. However, while perfectly feasible this restriction may lead to badly localized Wannier functions that require large supercells. In the present work states close to the Fermi energy are treated as frozen to ensure maximum accuracy. Fortunately the localization properties are barely affected and the supercells are very large in the relevant cases.

Conditioned minimization and penalty functionals: Another potential problem is caused by the fact that the above localization criterion in terms of the spread operator represents a *global* criterion. Especially in systems containing vacuum regions it may be highly favourable for some Wannier functions to wander away from the system. By moving into the vacuum regions they enhance the localization of the remaining Wannier functions. However, the resulting WF set is useless of course. This behaviour may be prevented by introducing a *penalty* functional to the total spread, i. e. by attaching a spring-like potential to the Wannier function centers

$$\Omega_P = A \sum_n w_n \left[\langle r \rangle_n - r_{n0} \right]^2, \qquad (2.201)$$

where A represents an arbitrarily chosen amplitude of the functional and r_{n0} denotes the target position of the n-th Wannier function.

2.2.4d Obtaining the real-space Hamiltonian

As illustrated in section 2.2.4a the actual conductance calculation requires only the Hamiltonian's matrix elements in real space within a localized orbital basis. In the basis set of Wannier functions the WF Hamiltonian $\mathcal{H}_{ij}(R) = \langle w_{i0} | \mathcal{H} | w_{jR} \rangle$ can be computed in a simple manner from the unitary matrices $U_{mn}^{(k)}$ obtained during the

2.2. Transport theory

localization procedure:

$$\mathcal{H}^{(rot)}(\mathbf{k}) = U^{(\mathbf{k})\dagger}\mathcal{H}(\mathbf{k})U^{(\mathbf{k})} \qquad (2.202)$$

Subsequently $\mathcal{H}^{(rot)}(\mathbf{k})$ is Fourier transformed into real space with the corresponding set of Bravais lattice vectors $\mathbf{R}$:

$$\mathcal{H}_{ij}(\mathbf{R}) = \frac{1}{N_k}\sum_{\mathbf{k}} e^{-i\mathbf{k}\cdot\mathbf{R}}\, \mathcal{H}_{ij}^{(rot)}(\mathbf{k}) \qquad (2.203)$$

Note that it should be tested whether the resulting Hamiltonian still describes the system correctly. This can be done by comparing the eigenvalues of this Hamiltonian to the eigenvalues obtained from DFT. Naturally, both bandstructures should be identical.

2.3 Optical and spectroscopic properties

The third section of the method review addresses the derivation of optical and spectroscopic properties from density functional calculations. First the concept of reflectance anisotropy spectroscopy (RAS) is introduced, which represents a highly versatile and widely-used tool for optical probing of surfaces. It is demonstrated how RA spectra can be derived from the surface dielectric function of the system. Subsequently the discussion proceeds to the calculation of dielectric functions from DFT. Since DFT is essentially a ground-state theory, while optical excitations are by definition excited state properties, Green's function techniques are presented to incorporate the relevant many-body aspects into the DFT treatment. The section closes with a short discussion of the implications of these many-body aspects on the band structure and dielectric function at the example of $LiNbO_3$.

2.3.1 Reflectance anisotropy spectroscopy (RAS)

Reflectance anisotropy spectroscopy (RAS) is a highly successful technique for the non-destructive in-situ optical probing of surfaces [133, 134]. It measures the relative reflectivity of a surface depending on the specimens spatial orientation and the wavelength. As it turns out many surface reconstructions feature unique RAS signatures. However, since it is a very indirect method accompanying calculations are a necessity. The basic theory for RAS was developed by Bagchi [140] and extended later by Del Sole [141]. It's first application in supercell slab calculations was demonstrated by Manghi [143]. In the present work it is employed to confirm or reject competing structural models by predicting RAS data for different surface reconstructions and subsequent comparison with experiment.

The surface contribution to the reflectance $\Delta R/R$ for s-light polarized along i and normal incidence can be expressed as [139, 140, 141]

$$\frac{\Delta R_i}{R}(\omega) = \frac{4\omega}{c} \Im \left\{ \frac{\Delta \epsilon_{ii}(\omega)}{\epsilon_b(\omega) - 1} \right\} \tag{3.204}$$

where R represents the reflectance according to the Fresnel equation, ϵ_b denotes the bulk dielectric function and $\Delta \epsilon_{ij}$ is given by:

$$\begin{aligned}\Delta \epsilon_{ij} &= \int dz\, dz' \, [\epsilon_{ij}(\omega; z, z') - \delta_{ij}\delta(z-z')\epsilon_0(\omega; z)] \\ &- \int dz\, dz'\, dz''\, dz''' \; \epsilon_{iz}(\omega; z, z') \epsilon_{zz}^{-1}(\omega; z', z'') \epsilon_{zj}(\omega; z'', z''')\end{aligned} \tag{3.205}$$

2.3. Optical and spectroscopic properties

$\epsilon_{ij}(\omega; z, z')$ denotes the non-local *macroscopic* dielectric tensor of the solid-vacuum interface that accounts for all many-body aspects and local field-effects [142]. The surface is located in the plane $z = 0$, with the bulk material underneath proceeding along $z > 0$. However, the second term is difficult to evaluate due to the inverse dielectric tensor and the 4-fold integration. In practical calculations it is usually neglected [143]. The outer integration of the first term may be stopped at $z = d$ if surface-induced effects have vanished at the depth d, with $\epsilon_{ij}(\omega; z, z') \approx \epsilon_b(\omega)\delta(z - z')\delta_{ij}$. Eq. (3.205) can then be evaluated numerically within the supercell approach. It is advisable to use a symmetric slab that features the surfaces of interest identically on both sides. This approach enlarges the supercell by introducing a slab thickness of $2d$, but avoids the necessity of truncating the resulting charge density and wavefunctions to remove the effects of unwanted surfaces, i. e., a hydrogen terminated bottom side of an asymmetric slab. The diagonal components $\Delta\epsilon_{ij}$ then assume the form

$$\Delta\epsilon_{ij}(\omega) = \frac{1}{2} \int dz \int dz' \, \epsilon_{ii}^{slab}(\omega; z, z') - 2d \cdot \epsilon_b(\omega) = d \left[\epsilon_{ii}^{slab}(\omega) - \epsilon_b(\omega) \right], \quad (3.206)$$

where the factor $1/2$ compensates for the fact that a symmetric slab features *two* identical surfaces. Provided (i) a sufficiently large crystal slab to describe both the surface and surface-modified bulk wavefunctions and (ii) sufficiently small off-diagonal terms of the dielectric tensor in comparison to the diagonal ones, $\Delta R/R$ can be expressed as [143]:

$$\boxed{\frac{\Delta R}{R}(\omega) = \frac{2\omega d}{c} \Im \left\{ \frac{\epsilon_{ii}^{slab}(\omega) - \epsilon_{jj}^{slab}(\omega)}{\epsilon_b(\omega) - 1} \right\}} \quad (3.207)$$

Given the two conditions mentioned above Eq. (3.207) contains in principle all surface contributions to the optical reflectance. How the required dielectric tensors ϵ_{ij}^{slab} can be obtained from *first principles* DFT ground-state calculations is the topic of the next sections.

2.3.2 Obtaining the dielectric tensor from DFT

Despite the known bandgap problem of the Kohn-Sham formalism optical calculations are routinely based upon the KS-eigenvalues due to the efficiency of the approach. Conceptually missing aspects, i. e., electron screening effects and electron-hole interaction, can be subsequently incorporated by Green's function techniques as required or computationally feasible. In fact a treatment at the KS-level of theory represents the only possible choice in many instances. The computational demands of such many-body Green's function techniques are still prohibitive for larger systems, even on today's fastest supercomputers.

2.3.2a   Independent particle approximation (IPA)

Calculating the dielectric tensor involves the response of a system's density $n(r)$ to an external time-varying potential $V_{ext}(r,t)$. The potential $V_{ext}(r,t)$ induces a linear response $n_{ind}(r,t)$ according to

$$n_{ind}(q+G;\omega) = \frac{1}{V}\sum_{G'} P(q+G,q+G';\omega) V_{ext}(q+G';\omega) \quad (3.208)$$

in reciprocal space, with

$$n_{ind}(q+G;\omega) = \frac{1}{\sqrt{2\pi V}} \int_{-\infty}^{\infty} dt \int d^3r\, e^{-i(q+G)r+i\omega t} n_{ind}(r,t) \quad (3.209)$$

$$V_{ext}(q+G;\omega) = \frac{1}{\sqrt{2\pi V}} \int_{-\infty}^{\infty} dt \int d^3r\, e^{-i(q+G)r+i\omega t} V_{ext}(r,t) \quad (3.210)$$

and the *polarisation function* given by:

$$P(q+G,q+G';\omega) = 2\sum_{n,k}\sum_{n',k'} B_{nn'}^{kk'}(q+G) B_{nn'}^{kk'}(q+G')$$
$$\cdot \frac{f_{n'}(k') - f_n(k)}{\epsilon_{n'}(k') - \epsilon_n(k) + \hbar(\omega + i\eta)} \quad (3.211)$$

$$B_{nn'}^{kk'}(q) = \langle \phi_{nk} | e^{iq\cdot r} | \phi_{n'k'} \rangle \quad (3.212)$$

The density n_{ind} in turn induces a screening potential V_{ind} that can be written as:

$$V_{ind}(q+G;\omega) = \frac{4\pi e^2}{\|q+G\|^2} n_{ind}(q+G;\omega) \quad (3.213)$$

These functions allow to define the dielectric tensor in reciprocal space according to:

$$V_{ext}(q+G;\omega) = \sum_{G'} \epsilon(q+G,q+G';\omega)[V_{ext}+V_{ind}](q+G';\omega) \quad (3.214)$$

By neglecting any non-diagonal elements of ϵ and thus local field effects one can obtain the dielectric function as:

2.3. Optical and spectroscopic properties

$$\epsilon(q+G;\omega) = 1 + \frac{8\pi e^2}{V\|q+G\|^2} \cdot \sum_{nk'}\sum_{n'k'} \left\| B_{nn'}^{kk'}(q+G) \right\|^2$$

$$\cdot \frac{f_n(k) - f'_n(k')}{\epsilon_{n'}(k') - \epsilon_n(k) + \hbar(\omega + i\eta)} \quad (3.215)$$

For $G = 0$ this expression is the *Ehrenreich-Cohen formula* [145]. Solving the KS-problem yields the required eigenenergies $\epsilon_n(k)$ and occupation numbers $f_n(k)$. In the optical limit the wavevector q converges towards $q \to 0$. This allows to calculate the Bloch integrals in Eq. (3.212) according to

$$\lim_{q\to 0} \langle \phi_{nk} | e^{iq\cdot r} | \phi_{n'k'} \rangle = \lim_{q\to 0} \frac{1}{\epsilon_n(k) - \epsilon_{n'}(k')} \langle \phi_{nk} | \left[\mathcal{H}, e^{iq\cdot r} \right] | \phi_{n'k'} \rangle$$

$$= \frac{1}{\epsilon_n(k) - \epsilon_{n'}(k')} \lim_{q\to 0} \langle \phi_{nk} | q \cdot v | \phi_{n'k'} \rangle \quad (3.216)$$

$$v_i = \frac{dr_i}{dt} = \frac{i}{\hbar}[\mathcal{H}, r_i] = \lim_{q_i \to 0} \frac{1}{\hbar q_i} \left[\mathcal{H}, e^{iq_i r_i} \right] \quad (3.217)$$

where v_i denotes the components of the velocity operator v. Inserting into Eq. (3.214) yields:

$$\epsilon(q;\omega) = \sum_{i,j} \tilde{q}_i \epsilon_{ij}(\omega) \tilde{q}_j, \quad \tilde{q}_k = \frac{q_k}{\|q\|} \quad (3.218)$$

$$\epsilon_{ij}(\omega) = \delta_{ij} + \frac{4\pi e^2 \hbar^2}{V} \sum_{nk}\sum_{nk'} \frac{\langle \phi_{nk} | v_i | \phi_{n'k'} \rangle \langle \phi_{n'k'} | v_j | \phi_{nk} \rangle}{[\epsilon_n(k) - \epsilon_{n'}(k')]^2}$$

$$\cdot \frac{f_n(k) - f'_n(k')}{\epsilon_{n'}(k') - \epsilon_n(k) + \hbar(\omega + i\eta)} \quad (3.219)$$

Employing the PAW-approach $\mathcal{H}$ represents the all-electron Hamiltonian and $|\phi_{nk}\rangle$ are the all-electron wavefunctions. In semiconducting or insulating cases at $T = 0$ Eq. (3.219) can be simplified even further [145, 146, 147]:

$$\boxed{\epsilon_{ij}(\omega) = \delta_{ij} + \frac{8\pi e^2 \hbar^2}{V m_e^2} \sum_{k}\sum_{v,c} \frac{\langle \phi_{vk} | p_i | \phi_{ck} \rangle \langle \phi_{ck} | p_j | \phi_{vk} \rangle}{[\epsilon_c(k) - \epsilon_v(k)]([\epsilon_c(k) - \epsilon_v(k)]^2 - \hbar[\omega + i\eta]^2)}} \quad (3.220)$$

The indices c, v traverse through the conduction and valence bands, respectively.

2.3.2b Intraband contributions in metallic case

Although it is a classical model, the low energy optical properties of simple metals can be interpreted in terms of the Drude model for the free electron gas and the propagation of electromagnetic waves through a medium. The Maxwell equations for a non-magnetic medium in the absence of any external charges or currents but presence of the internal charge density ρ_{int} and current density J_{int} are given in Gauss units by [148]:

$$\boxed{\begin{array}{ll} \text{div} E = 4\pi \rho_{int} & \text{rot} E = -\frac{1}{c}\frac{\partial B}{\partial t} \\ \text{div} B = 0 & \text{rot} B = \frac{1}{c}\frac{\partial E}{\partial t} + \frac{4\pi}{c} J_{int} \end{array}} \qquad (3.221)$$

Consider the propagation of electromagnetic waves through a homogeneous medium with the following geometric setup:

$$E(r,t) = E(z)e^{-i\omega t}(1,0,0) \qquad (3.222)$$
$$B(r,t) = B(z)e^{-i\omega t}(0,1,0) \qquad (3.223)$$
$$J_{int}(r,t) = J_{int}(z)e^{-i\omega t}(1,0,0) \qquad (3.224)$$

As a consequence of Eq. (3.222) $\text{div} E = 0$, which combined with the first Maxwell equation implies $\rho_{int} = 0$. Eq. (3.223) is consistent with the second Maxwell equation $\text{div} B = 0$. The remaining two Maxwell equations yield:

$$\frac{dE(z)}{dz} = \frac{i\omega}{c} B(z), \quad -\frac{dB(z)}{dz} = -\frac{i\omega}{c} E(z) + \frac{4\pi}{c} J_{int}(z) \qquad (3.225)$$

By eliminating $B(z)$ one obtains:

$$\boxed{\frac{d^2 E(z)}{dz^2} = -\frac{\omega^2}{c^2} E(z) - \frac{4\pi i \omega}{c^2} J_{int}(z)} \qquad (3.226)$$

The current density is linked to the electric field by the common approximation

$$J_{int}(z) = \sigma(\omega) E(z) \qquad (3.227)$$

where $\sigma(\omega)$ is the so-called *conductivity function* of the medium. Inserting Eq. (3.227) into Eq. (3.226) yields:

$$\frac{d^2 E(z)}{dz^2} = -\frac{\omega^2}{c^2}\left[1 + \frac{4\pi i \omega \sigma(\omega)}{\omega}\right] E(z) \qquad (3.228)$$

This equation is solved by a damped wave of the form $E(z) = E_0 e^{i\frac{\omega}{c} N z}$ where the

2.3. Optical and spectroscopic properties

complex refractive index $N(\omega)$ is given by:

$$\boxed{N^2(\omega) = \epsilon(\omega) = 1 + \frac{4\pi i \sigma(\omega)}{\omega}} \quad (3.229)$$

Thus the dielectric function of the system can be obtained from the complex conductivity function $\sigma(\omega)$, which is easily derived for the Drude model. Consider a free-electron gas with n carriers per unit volume, each with an effective mass m and charge $-e$. The electrons are embedded in a uniform background of neutralizing positive charge. Then the classical equation of motion for an electron in the presence of an electric field $E(r,t) = E_0 e^{i(q \cdot r - \omega t)}$ of wavevector q and frequency ω is given by [148]

$$m\ddot{r} = -\frac{m}{\tau}\dot{r} - eE_0 e^{i(q \cdot r - \omega t)} \quad (3.230)$$

where $r(t)$ denotes the coordinate of the electron. τ represents a phenomenological relaxation time, controlling the viscous damping term $-(m/\tau)\dot{r}$. This damping term is meant to account for the various dissipation mechanisms such as random collisions between the electrons, defect- and impurity-scattering and electron-phonon interaction. Assuming the spatial excursions of $r(t)$ around any point r_0 to be much smaller than the driving field's wavelength, one can replace $r(t)$ in the exponent by r_0. Without loss of generality one may assume $r_0 = 0$ and obtains:

$$m\ddot{r} = -\frac{m}{\tau}\dot{r} - eE_0 e^{-i\omega t} \quad (3.231)$$

Inserting $r(t)$ in the form $r(t) = A_0 e^{-i\omega t}$ yields:

$$A_0 = \frac{e\tau}{m}\frac{1}{\omega(i+\omega\tau)}E_0 \quad (3.232)$$

The free-carrier contribution to the current density is given by:

$$J = -ne\dot{r} = nei\omega A_0 e^{-i\omega t} = \frac{ne^2\tau}{m}\frac{1}{1-i\omega\tau}E_0 e^{-i\omega t} \quad (3.233)$$

Thus the frequency dependent complex conductivity can be derived as:

$$\sigma(\omega) = \frac{ne^2\tau}{m}\frac{1}{1-i\omega\tau} = \sigma_0 \frac{1}{1-i\omega\tau} \quad (3.234)$$

Note that the Drude theory neglects the wavevector dependence of the electric field and optical constants (also known as *spatial dispersion*) and provides only $\sigma(q \to 0, \omega)$.

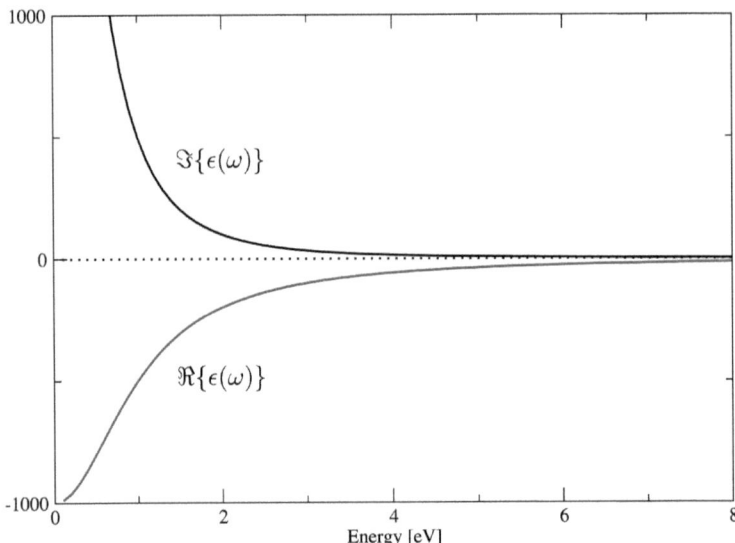

Figure 3.9: *Schematic representation of the Drude free electron gas dielectric function ($\omega_p^2 = 10^4 s^{-1}, \tau = 1s$).*

According to Eq. (3.229) and (3.234) the dielectric constant becomes

$$\epsilon(\omega) = 1 - \frac{\omega_p^2}{\omega(\omega + i/\tau)} \quad (3.235)$$

$$\Re\{\epsilon(\omega)\} = 1 - \frac{\omega_p^2 \tau^2}{1 + \omega^2 \tau^2}, \quad \Im\{\epsilon(\omega)\} = \frac{\omega_p^2 \tau}{\omega(1 + \omega^2 \tau^2)} \quad (3.236)$$

where ω_p denotes the plasma frequency of the respective material. In this case for the free electron gas ω_p is given by $\omega_p^2 = (4\pi n e^2)/m$. A schematic representation of the dielectric function for the Drude free electron gas is shown in Fig. 3.9.

Returning to a quantum mechanical point of view it is no longer sufficient to incorporate only *interband* transitions between valence and conduction bands as in Eq. (3.220). A metal features partially occupied bands that cross the Fermi energy. Thus beside the interband transitions a metal also exhibits *intraband* transitions. These dominate at low energies where few to none interband transitions are available anymore. They are responsible for the low energy characteristics of the dielectric function in analogy to the Drude model. In reciprocal space neglecting local field effects the intraband

2.3. Optical and spectroscopic properties

contribution to the dielectric tensor can be expressed as [151]

$$\epsilon_{intra}(\omega) = 1 - \lim_{q \to 0} \left\{ \frac{4\pi}{\|q\|^2} \sum_n \int_{BZ} d^3k \, \frac{1}{(2\pi)^3} [f_n(k-q) - f_n(k)] \right. \\ \left. \times \frac{\|\langle nk|e^{iq \cdot r}|nk-q\rangle\|^2}{w + \epsilon_n(k-q) - \epsilon_n(k)} \right\} \quad (3.237)$$

with the sum n restricted to partially occupied bands only. The relation

$$f_n(k-q) - f_n(k) = [f_n(k-q) - f_n(k)] \\ \cdot \{\theta[f_n(k-q) - f_n(k)] - \theta[f_n(k) - f_n(k-q)]\} \quad (3.238)$$

and employing time-reversal symmetry allows to rewrite Eq. (3.237) as:

$$\epsilon_{intra}(\omega) = 1 - \lim_{q \to 0} \left\{ \frac{8\pi}{\|q\|^2} \sum_n \int_{BZ} \frac{d^3k}{(2\pi)^3} \right. \\ \cdot [f_n(k-q) - f_n(k)] \cdot \theta[f_n(k-q) - f_n(k)] \\ \left. \times \frac{\|\langle nk|e^{iq \cdot r}|nk-q\rangle\|^2 [\epsilon_n(k) - \epsilon_n(k-q)]}{\omega^2 - [\epsilon_n(k) - \epsilon_n(k-q)]^2} \right\} \quad (3.239)$$

In the limit $q \to 0$ Eq. (3.239) yields the Drude contribution to the dielectric function:

$$\epsilon_{intra}(\omega) = 1 - \frac{\omega_p^2}{\omega^2} + \mathcal{O}(\|q\|^2) \quad (3.240)$$

$$\omega_p^2 = \lim_{q \to 0} \left\{ \frac{8\pi}{\|q\|^2} \sum_n \int_{BZ} \frac{d^3k}{(2\pi)^3} \right. \\ \cdot [f_n(k-q) - f_n(k)] \cdot \theta[f_n(k-q) - f_n(k)] \\ \left. \times \|\langle nk|e^{iq \cdot r}|nk-q\rangle\|^2 [\epsilon_n(k) - \epsilon_n(k-q)] \right\} \quad (3.241)$$

The θ-function in Eq. (3.241) strongly limits the region of the Brillouin zone that contributes to the integral. Hence the k-point mesh and the chosen q-vector become two very critical convergence parameters in the numerical evaluation of Eq. (3.241). For small q-vectors only very few k-points satisfy the condition $[f_n(k-q) - f_n(k)] \neq 0$. However, in practical calculations q must be small enough to ensure the limit $q \to 0$, while at the same time a sufficiently large number of k-points must still contribute to the sum. Hence a k-point mesh spacing no larger than $k_{n+1} - k_n \leq q$ is enforced by this condition. To attain convergence an enormous number of k-points is required, as will be seen in the later application of this scheme.

2.3.3 Many-body correction Green's function schemes

Optical properties are by definition characteristics of excited states. However, density functional theory is essentially a ground-state theory and thus describes optical properties inaccurately in many instances. The most important shortcomings are:

- Disregard of *screened* electron-electron interaction resulting in a severe underestimation of band gaps.

- Missing *electron-hole interaction* for excited states leading to a major redistribution of the spectral weights.

Green's function techniques are highly successful for incorporating these many-body aspects of excited states. They allow for a systematic improvement of observables connected to electronic excitations by providing a well defined perturbation series. In the following sections two approaches will be presented for the perturbative treatment of one-electron excitations and electron-hole interaction.

2.3.3a One- and two-particle Green's functions

The one-electron Green's function of a N-electron system is defined as [155]

$$\mathcal{G}(rt, r't') = \langle N | \mathcal{T} \{ \psi(r,t) \psi^\dagger(r',t') \} | N \rangle \qquad (3.242)$$

where $\psi^\dagger(r,t)$ ($\psi(r,t)$) is a field operator in the Heisenberg representation which creates (annihilates) an electron at (r,t), respectively. $|N\rangle$ represents the exact N-electron ground-state and $\mathcal{T}$ denotes the time-ordered product of $\psi, \psi^\dagger$. Depending on t, t' the Green's function $\mathcal{G}(rt, r't')$ may describe two different processes:

$t < t'$: $\mathcal{G}(rt, r't')$ describes a $(N-1)$ electron-state, where a hole created at (r,t) propagates to (r',t').

$t > t'$: $\mathcal{G}(rt, r't')$ describes a $(N+1)$ electron-state, where an extra electron created at (r,t) propagates to (r',t').

The one-particle Green's function allows to describe one-electron excitation processes including screening effects, such as, i. e., photoelectron spectroscopy. In analogy the two-particle Green's function is defined according to

$$\mathcal{G}(1,2;1',2') = \langle N | \mathcal{T} \{ \psi(1) \psi(2) \psi^\dagger(1') \psi^\dagger(2') \} | N \rangle \qquad (3.243)$$

with the abbreviated notation $n \equiv (r_n, t_n)$. Depending on the time ordering the two-particle Green's function represents the following propagators:

2.3. Optical and spectroscopic properties

$t_1, t_2 < t'_1, t'_2$: hole-hole propagation

$t_1, t_2 > t'_1, t'_2$: electron-electron propagation

$t_1, t'_1 \lessgtr t_2, t'_2$: electron-hole propagation

Electron-hole propagation is the most relevant process regarding optical excitations. It is examined in more detail in section 2.3.3d. From the Heisenberg equation of motion for the field operator $i\partial\psi(r)/\partial t = [\psi(r), \mathcal{H}]$ one can obtain the equation of motion for the one-particle Green's function:

$$\left[i\frac{\partial}{\partial t} + \frac{1}{2}\nabla^2 - V(r)\right]\mathcal{G}(rt, r', t')$$
$$+ i\int d^3r''\, v(r, r'')\langle N|\mathcal{T}\{\psi(r,t)\psi(r'',t)\psi^\dagger(r'',t)\psi^\dagger(r',t')\}|N\rangle \qquad (3.244)$$
$$= \delta(r - r')\delta(t - t')$$

The solution to Eq. (3.244) obviously requires knowledge of the two-particle Green's function, which results from the electron-electron Coulomb interaction terms. In an analogous way one could now construct an equation of motion for the two-particle Green's function, repeating the process recursively. However, it is much more convenient to avoid the necessity of calculating the two-particle Green's function at all. This can be achieved by introducing the concept of the *electronic self-energy* as the central quantity.

2.3.3b Electronic self-energy and Hedin's equations

The electronic self-energy Σ is introduced by rewriting Eq. (3.244) according to [155]

$$\left[i\frac{\partial}{\partial t} + \frac{1}{2}\nabla^2 - V(r) - V_H(r)\right]\mathcal{G}(rt, r', t')$$
$$- i\int d^3r''\, dt''\, \Sigma(rt, r'', t'')\mathcal{G}(r''t'', r', t') \qquad (3.245)$$
$$= \delta(r - r')\delta(t - t')$$

where the Hartree potential V_H was extracted from the Coulomb integral in Eq. (3.244). While this approach does not require obtaining the two-particle Green's function anymore, an explicit expression for Σ has yet to be determined. This is achieved by deriving a set of five integro-differential equations, the so-called *Hedin equations* [153, 154], that eventually define the self-energy. Fourier transforming Eq. (3.245) into frequency space yields:

$$\left[\omega + \frac{1}{2}\nabla^2 - V(r) - V_H(r)\right] \mathcal{G}(r,r',\omega) - \int d^3r'' \, \Sigma(r,r'',\omega) \mathcal{G}(r'',r',\omega)$$
$$= \delta(r - r') \tag{3.246}$$

Defining $\mathcal{G}_0$ as the Green's function corresponding to $\Sigma = 0$ allows to rewrite Eq. (3.246) in the form of a Dyson equation

$$\boxed{\mathcal{G}(1,2) = \mathcal{G}_0(1,2) + \int d3 \, d4 \, \mathcal{G}_0(1,3) \Sigma(3,4) \mathcal{G}(4,2)} \tag{3.247}$$

where the abbreviated notation $n \equiv (r_n, t_n)$ is reintroduced. This is already one of the five Hedin equations to be derived. Several ways exist for the derivation of an explicit expression for the self-energy. Martin and Schwinger [156, 157] suggested employing a time-varying field $\phi(r,t)$ as a mathematical tool for evaluating the self-energy and setting it to 0 once the self-energy is obtained. The functional derivative of $\mathcal{G}$ with respect to ϕ is given by:

$$\frac{\partial \mathcal{G}(1,2;\phi)}{\partial \phi(3^+)} = \mathcal{G}(1,2;\phi)\mathcal{G}(3,3^+;\phi) - \mathcal{G}(1,2;3,3^+;\phi) \tag{3.248}$$

The superscript "+" indicates an infinitesimal imaginary shift of t to ensure convergence. Together with Eq. (3.243) this relation allows to replace the two-particle Green's function in Eq. (3.244). Comparing the result with Eq. (3.245), subsequently applying the identity

$$\frac{\partial}{\partial \phi}(\mathcal{G}^{-1}\mathcal{G}) = \mathcal{G}^{-1}\frac{\partial \mathcal{G}}{\partial \phi} + \frac{\mathcal{G}^{-1}}{\partial \phi}\mathcal{G} = 0 \to \frac{\partial \mathcal{G}}{\partial \phi} = -\mathcal{G}\frac{\mathcal{G}^{-1}}{\partial \phi}\mathcal{G} \tag{3.249}$$

and evaluating $\partial \mathcal{G}^{-1}/\partial \phi$, where from Eq. (3.245) $\mathcal{G}^{-1}$ is given by

$$\mathcal{G}^{-1} = i\frac{\partial}{\partial t} + \frac{1}{2}\nabla^2 - V(r) - V_H(r) - \Sigma \tag{3.250}$$

finally yields the desired expression for the self-energy:

$$\boxed{\Sigma(1,2) = i \int d3 \, d4 \, \mathcal{G}(1,3^+) W(1,4) \Lambda(3,2,4)} \tag{3.251}$$

2.3. Optical and spectroscopic properties

Λ represents the *vertex function*

$$\Lambda(1,2,3) = -\frac{\partial \mathcal{G}^{-1}(1,2)}{\partial V(3)} \tag{3.252}$$

$$= \delta(1-2)\delta(2-3) + \frac{\partial \Sigma(1,2)}{\partial V(3)} \tag{3.253}$$

$$= \delta(1-2)\delta(2-3)$$
$$+ \int d4\, d5\, d6\, d7\, \frac{\partial \Sigma(1,2)}{\partial \mathcal{G}(4,5)} \mathcal{G}(4,6)\mathcal{G}(7,5)\Lambda(6,7,3) \tag{3.254}$$

where the second line is obtained from Eq. (3.251), the last line by applying the chain rule $\partial \Sigma / \partial V = (\partial \Sigma / \partial \mathcal{G})(\partial \mathcal{G} / \partial V)$ and utilizing the identity Eq. (3.249). W represents the *screened Coulomb potential* according to:

$$W(1,2) = \int d3\, v(3-2)\epsilon^{-1}(1,3) \tag{3.255}$$

$$\epsilon^{-1}(1,2) = \frac{\partial V(1)}{\partial \phi(2)} = 1 + v\frac{\partial n}{\partial \phi} = 1 + v\frac{\partial n}{\partial V}\frac{\partial V}{\partial \phi} \tag{3.256}$$

The *response* and *polarization functions* are defined, respectively, as:

$$R(1,2) = \frac{\partial n(1)}{\partial \phi(2)}, \quad P(1,2) = \frac{\partial n(1)}{\partial V(2)} \tag{3.257}$$

R represents the change in the charge density n in response to the change in the *external* field, while P returns the change in n due to a change in the *total* external+induced field. The relation $n(1) = -i\mathcal{G}(1,1^+)$ allows to rewrite P as:

$$P(1,2) = -i \int d3\, d4\, \mathcal{G}(1,3)\Lambda(3,4,2)\mathcal{G}(4,1^+) \tag{3.258}$$

Concerning the screened Coulomb interaction W follows as:

$$\epsilon^{-1} = 1 + vR, \ \epsilon = 1 - vP \Rightarrow R = P + PvR$$
$$\Rightarrow \boxed{W = v + vPW = v + vRv} \tag{3.259}$$

Summarizing the previous results one obtains the 5 coupled integro-differential *Hedin equations* [153, 154] according to:

$$\Sigma(1,2) = i \int d3\, d4\, \mathcal{G}(1,3^+)W(1,4)\Lambda(3,2,4) \tag{3.260}$$

$$W(1,2) = v(1,2) + \int d3\, d4\, v(1,3)P(3,4)W(4,2) \tag{3.261}$$

$$P(1,2) = -i \int d3\, d4\, \mathcal{G}(1,3)\Lambda(3,4,2)\mathcal{G}(4,1^+) \tag{3.262}$$

$$\Lambda(1,2,3) = \delta(1-2)\delta(2-3)$$
$$+ \int d4\, d5\, d6\, d7\, \frac{\partial \Sigma(1,2)}{\partial \mathcal{G}(4,5)} \mathcal{G}(4,6)\mathcal{G}(7,5)\Lambda(6,7,3) \tag{3.263}$$

$$\mathcal{G}(1,2) = \mathcal{G}_0(1,2) + \int d3\, d4\, \mathcal{G}_0(1,3)\Sigma(3,4)\mathcal{G}(4,2) \tag{3.264}$$

It should be noted that this set of coupled equations is formally exact without any approximations and in principle allows access to all one-electron properties of the system. The dynamically screened Coulomb interaction W is central to the calculation of the self-energy. W in turn is determined by the polarization P, which describes the N-electron system's response with respect to an additional electron or hole. P and Σ are both governed by the *vertex function* Λ that describes the multitude of interactions ocurring between screening electrons and holes. However, to utilize Hedin's equations in practical applications one needs to derive a calculational scheme for solving this system of coupled equations. The solution is usually acquired within the so-called *independent quasi-particle approximation*.

2.3.3c Independent quasi-particle approximation (IQA)

Assuming $\Sigma = 0$ the vertex function is given by $\Lambda(1,2,3) = \delta(1-2)\delta(2-3)$. Inserting this vertex function into the Hedin equations yields:

$$\Sigma(1,2) = i\mathcal{G}(1,2)W(1^+,2) \tag{3.265}$$

$$W(1,2) = v(1,2) + \int d3\, d4\, v(1,3)P(3,4)W(4,2) \tag{3.266}$$

$$P(1,2) = -i\mathcal{G}(1,2)\mathcal{G}(2,1) \tag{3.267}$$

This specific approximation is named *GW-approximation* (GWA) since Σ is given by the product of the one-electron Green's function $\mathcal{G}$ and the screened Coulomb interaction W. On first glance this equation set seems well suited for an iterative self-consistent solution algorithm. By starting from a suitable Green's function $\mathcal{G}_0(12)$ a polarization function $P_0 = -i\mathcal{G}_0(12)\mathcal{G}_0(21)$ can be obtained. Subsequently the self-energy $\Sigma_0 = \mathcal{G}_0 W_0$ is calculated, with $W_0 = \epsilon_0^{-1} v$. Afterwards the Green's function is updated according to the Dyson equation $\mathcal{G} = \mathcal{G}_0 + \mathcal{G}_0(\Sigma - V_{eff})\mathcal{G}$ and the self-consistency

2.3. Optical and spectroscopic properties

loop can be closed. However, experience with these types of calculations show that the first iteration $\Sigma = \mathcal{G}_0 W_0$ already gives results in good agreement with experiment. Further iterations even reduce the accuracy of the results in many cases. This can be qualitatively understood by taking a look back at Eq. (3.245). From this equation it is immediately obvious that $\Sigma = 0$ corresponds to the *Hartree approximation*. Thus the Green's function in the GW-approximation represents in fact the *Hartree Green's function*. Therefore $\mathcal{G}$ cannot be expected to converge towards the exact Green's function from the original Hedin equations. It should also be remembered that the vertex function Λ remains static in this scheme and is not updated accordingly [158]. For these reasons the common approach is to employ the $G_0 W_0$-approximation in practical calculations.

With regard to DFT the GW-approximation can be implemented by post-processing the DFT results. The GW quasi-particle energies are obtained from the DFT eigenvalues employing the following perturbative approach [135]

$$\epsilon_n(k)^{QP} = \epsilon_n(k) + \frac{1}{1+\beta_{n,k}} \left[\Sigma_{n,k}^{stat} + \Sigma_{n,k}^{dyn}(\epsilon_n(k)) - V_{n,k}^{XC} \right] \quad (3.268)$$

where the self-energy operator Σ has been divided into static and dynamic contributions, respectively. $\beta_{n,k}$ denotes the linear coefficient in the expansion of Σ^{dyn} around the DFT eigenvalue $\epsilon_n(k)$. The static part can be split into two constituents

$$\Sigma^{stat}(r,r') = \frac{1}{2} \sum_{n,k} \psi_{n,k}(r) \psi_{n,k}^{ast}(r') \left[W(r,r';0) - v(r-r') \right] \quad (3.269)$$

$$- \sum_{v,k} \psi_{v,k}(r) \psi_{v,k}^*(r') W(r,r';0) \quad (3.270)$$

representing the Coulomb hole Σ^{COH} and the screened exchange Σ^{SEX}. The summation for Σ^{SEX} is performed over the valence states v only. The Kohn-Sham wavefunctions $\psi_{n,k}$ are utilized as a starting point to construct $\mathcal{G}_0$. However, a major bottleneck is the calculation of the screened interaction W, because the dielectric tensor ϵ has to be inverted at every frequency ω. A tremendous speedup can be achieved by replacing the dielectric tensor by a model dielectric function, i. e. within the plasmon-pole approximation [160]. In the present work the version suggested by Bechstedt et al. [162] is utilized, where ϵ is given by

$$\epsilon(q,n) = 1 + \left[(\epsilon_\infty - 1)^{-1} + \frac{q^2}{q_{TF}(n)^2} + \frac{3q^4}{4k_F^2(n)q_{TF}^2(n)} \right]^{-1} \quad (3.271)$$

with k_F and q_{TF} representing the Fermi and Thomas-Fermi wavevectors, respectively. In analogy to the local density approximation Hybertsen and Louie [161] suggested to approximate the spatial dependence of the screening of the inhomogeneous system

$$W(r,r';0) = \frac{1}{2}\left[W^h(r-r',n(r)) + W^h(r-r',n(r)')\right] \quad (3.272)$$

by that of the homogeneous electron gas W^h. Eqs. (3.271) and (3.272) allow to derive Σ^{COH} analytically as

$$\Sigma^{COH}(r) = -\frac{q_{TF}(r)}{2}\sqrt{1 - \frac{1}{\epsilon_\infty}\left[1 + \frac{q_{TF}(r)}{k_F(r)}\sqrt{\frac{3\epsilon_\infty}{\epsilon_\infty - 1}}\right]} \quad (3.273)$$

where k_F and q_{TF} are evaluated at the local density $n(r)$. The matrix elements $\Sigma^{stat}_{n,k}$ are calculated in reciprocal space. However, during the Fourier transform of W only the diagonal elements are retained. Its nonlocality is approximated employing state-averaged electron densities

$$n_{n,k} = \int d^3r\, n(r)\|\psi_{n,k}(r)\|^2 \quad (3.274)$$

during the calculation of k_F and q_{TF}. The dynamic terms $\beta_{n,k}(r), \Sigma^{dyn}$ are approximated by integrals of the dielectric function (3.271) employing a single plasmon-pole to describe the frequency dependence. Again local-field effects are included by Eq. (3.274).

Scissors shift: While the GWA as described above is highly efficient, calculating quasi-particle shifts is still computationally very demanding. On the other hand, calculating the dielectric function requires a large number of k-points to attain convergence, regularly ranging between several hundreds up to thousands. Even on the fastest supercomputers available today the requirements for calculating the self-energy at every k-point can quickly exceed any reasonable amount of time. Thus a common approach is to calculate quasi-particle shifts only at a few selected k-points and examine the dependence of the shifts on the k-vector.

In many instances it turns out that the shifts depend rather little on the actual k-point but more on the distance of the DFT eigenvalue from the Fermi-energy and the orbital character of the states. I. e., surface states with a different orbital character than the bulk states often feature a somewhat smaller associated quasi-particle shift. In case the k-point dependence is sufficiently weak one can extrapolate an analyti-

2.3. Optical and spectroscopic properties

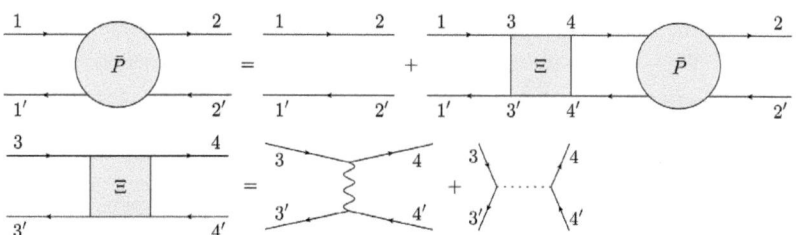

Figure 3.10: *Diagrammatic representation of the Bethe-Salpeter equation. The macroscopic polarization is related to the polarization of independent particles (1 → 2, 1' → 2') and an interaction kernel Ξ containing the screened electron-hole attraction (wriggled line) and the unscreened electron-hole exchange (dashed line) [135].*

cal functional from the selected k-points that returns the approximate quasi-particle shifts at any other. The shifts introduced by this approach are called *scissors shifts*. If performed carefully this approach yields almost exactly the same dielectric function as employing GW at every k-point, albeit at a significantly lower computational cost. This allows for treating larger systems which are otherwise inaccessible for today's technology.

2.3.3d Electron-hole attraction and Bethe-Salpeter Eq. (BSE)

Excitation energies obtained within the independent quasi-particle formalism describe only one-particle excitations, such as photoemission experiments. However, for the description of optical absorption the single quasi-particle formalism is insufficient. To improve the polarization function beyond the independent quasi-particle approximation it is necessary first to introduce a *two-particle* polarization function according to [135]:

$$P(1,2,3,4) = -\frac{i}{\hbar}\mathcal{G}(1,4)\mathcal{G}(3,2) \tag{3.275}$$

Inspecting the Hedin equations again reveals that one can obtain the two-particle polarization function by substituting first Eq. (3.263) into (3.275), then employ the relation $\partial \Sigma / \partial \mathcal{G} = W$ and subsequently insert Eq. (3.261). In GW-approximation with $\Lambda(1,2,3) = \delta(1-2)\delta(2-3)$ the polarization function assumes the form

$$\boxed{P(1,1,2,2) = P_0(1,1,2,2) - \hbar \int d3\, d4\, P_0(1,1,4,3) W(3,4) P(3,4,2,2)} \tag{3.276}$$

where P_0 represents the polarization function in independent quasi-particle approximation. This is the so-called *Bethe-Salpeter Equation* (BSE). The two-particle polarization function obtained by solving the BSE incorporates not only the quasi-particle character of the electrons in GWA, but the excitonic effect of electron-hole-attraction as well (cf. Fig. 3.10). With regard to DFT the Bethe-Salpeter equation (3.276) can be conveniently solved in the basis of Bloch functions defined by the Kohn-Sham formalism. Its solution can be expressed in resolvent representation as

$$P_{(n_1,n_2)(n_3,n_4)} = [\mathcal{H} - \omega]^{-1}_{(n_1,n_2)(n_3,n_4)} (f_{n_4} - f_{n_3}) \qquad (3.277)$$

with the two-particle Hamiltonian given by:

$$\begin{aligned}\mathcal{H}_{(n_1,n_2)(n_3,n_4)} &\equiv (\epsilon^{QP}_{n_1} - \epsilon^{QP}_{n_2})\delta_{(n_1,n_3)}\delta_{(n_2,n_4)} + (f_{n_2} - f_{n_1}) \\ &\times \int dr_1\, dr_2\, dr_3\, dr_4\, \psi_{n_1}(r_1)\psi^*_{n_2}(r_2)\psi^*_{n_3}(r_3)\psi_{n_4}(r_4) \\ &\times [\delta(r_1 - r_2)\delta(r_3 - r_4)\bar{v}(r_1 - r_3) \\ &\quad - \delta(r_1 - r_3)\delta(r_2 - r_4)W(r_1, r_2)] \end{aligned} \qquad (3.278)$$

$f_n = 0, 1$ is the occupation number of the state n, denoting both band index and wave vector. In principle one could now obtain the polarization from Eq. (3.277) by inverting this Hamiltonian at any desired frequency ω. However, this would be computationally far too expensive for any practical calculation due to the non-Hermiticity and large dimension of $\mathcal{H}$. Fortunately $\dim(\mathcal{H})$ can be reduced by a factor of 2 since only pairs containing one filled and one empty state contribute to the macroscopic polarization due to the factors $(f_{n_4} - f_{n_3})$ and $(f_{n_2} - f_{n_1})$ in Eqs. (3.277) and (3.278). Another reduction by a factor 2 can be achieved by neglecting the off-diagonal blocks that couple the Hermitian resonant part of $\mathcal{H}$ to the antiresonant part. By restricting the calculation to spin singlets, static screening, transitions without momentum transfer by photons and neglecting umklapp processes the excitonic Hamiltonian can be written in reciprocal space according to:

$$\begin{aligned}\mathcal{H}^{res}_{vck,v'c'k'} &= (\epsilon^{QP}_{ck} - \epsilon^{QP}_{vk})\delta_{vv'}\delta_{cc'}\delta_{kk'} + \frac{4\pi}{\Omega}\sum_{G,G'}\left\{2\frac{\delta_{GG'}(1 - \delta_{G0})}{\|G\|^2}B^{kk'}_{cv}(G)B^{k'k'*}_{c'v'}(G)\right. \\ &\quad \left. -\frac{\epsilon^{-1}(k - k' + G, k - k' + G', 0)}{\|k - k' + G\|^2}B^{kk'}_{cc'}(G)B^{kk'*}_{vv'}(G')\right\} \end{aligned} \qquad (3.279)$$

$$B^{kk'}_{nn'}(G) = \frac{1}{\Omega}\int dr\, u^*_{nk}(G)e^{iG\cdot r}u_{n'k'}(r) \qquad (3.280)$$

$B^{kk'}_{nn'}$ denotes the Bloch integral over the periodic parts u of the Bloch wavefunctions and Ω represents the unit cell volume. The numerical evaluation of Eq. (3.279) is still computationally very demanding due to the rank of $\mathcal{H}$ and the double sum over G, G'. Again, the computational cost can be significantly reduced by replacing the inverse

2.3. Optical and spectroscopic properties

dielectric function by the model dielectric function Eq. (3.271), which has previously been used for calculating the self-energy operator. However, even after calculating the Hamiltonian the straight forward evaluation of Eq. (3.277) is still computationally prohibitive due to the large number of pair states. The usual approach is to convert the problem of calculating the resolvent into a generalized eigenvalue problem, which can be solved by diagonalization [167, 168]. Employing the spectral representation

$$[\mathcal{H} - \omega]^{-1} = \sum_{\lambda,\lambda'} \frac{|A^\lambda\rangle S^{-1}_{\lambda,\lambda'} \langle A^{\lambda'}|}{E_\lambda - \omega} \tag{3.281}$$

with $|A^\lambda\rangle$ and E_λ denoting the eigenvectors and eigenvalues of the excitonic Hamiltonian

$$\mathcal{H}|A^\lambda\rangle = E_\lambda|A^\lambda\rangle, \quad S_{\lambda,\lambda'} = \langle A^{\lambda'}|A^\lambda\rangle \tag{3.282}$$

the diagonal components of the macroscopic polarizability are given by:

$$\alpha^M_{jj}(\omega) = \frac{4e^2\hbar^2}{\Omega} \sum_\lambda \left\| \sum_k \sum_{c,v} \frac{\langle ck|v_j|vk\rangle}{\epsilon_c(k) - \epsilon_v(k)} A^\lambda_{vck} \right\|^2$$
$$\times \left\{ \frac{1}{E_\lambda - \hbar(\omega + i\gamma)} + \frac{1}{E_\lambda + \hbar(\omega + i\gamma)} \right\} \tag{3.283}$$

v_j denotes the corresponding Cartesian component of the single-particle velocity operator and γ the damping constant. Here the contributions of the anti-resonant part of the exciton Hamiltonian is formally included, while the coupling parts are neglected. This expression is straight forward to calculate but requires the solution of the eigenvalue problem (3.282). Since $\dim(N) = N_v N_c N_k$ this algorithm scales with $\mathcal{O}(N^3)$, which is computationally prohibitive for larger systems. A better scaling approach is to formulate the calculation of the polarizability as an initial value problem. Introducing the dipole moment vectors

$$\mu^j_{vck} = \frac{\langle ck|v_j|vk\rangle}{\epsilon_c(k) - \epsilon_v(k)} \tag{3.284}$$

allows to rewrite Eq. (3.283) according to:

$$\alpha^M_{jj}(\omega) = \frac{4e^2\hbar^2}{\Omega} \sum_\lambda \|\langle \mu^j | A^\lambda \rangle\|^2 \left\{ \frac{1}{E_\lambda - \hbar(\omega + i\gamma)} + \frac{1}{E_\lambda + \hbar(\omega + i\gamma)} \right\} \tag{3.285}$$

This equation reads in Fourier representation

$$\alpha^M_{jj}(\omega) = \frac{4e^2\hbar^2}{\Omega} i \int_0^\infty dt \, e^{i(\omega + i\gamma)t} \{\langle \mu^j | \xi^j(t) \rangle - \langle \mu^j | \xi^j(t) \rangle^*\} \tag{3.286}$$

with the time evolution of the vector $|\xi^j(t)\rangle$ being driven by the pair Hamiltonian:

$$i\hbar \frac{d}{dt}|\xi^j(t)\rangle = \mathcal{H}|\xi^j(t)\rangle, \quad |\xi^j(0)\rangle = |\mu^j\rangle \qquad (3.287)$$

Subsequently the initial value problem defined by Eq. (3.287) is solved using a finite difference method [180] according to:

$$\mathcal{H}|\xi(t_{i+1})\rangle = i\hbar \frac{|\xi(t_{i+2})\rangle - |\xi(t_i)\rangle}{2\Delta t} \qquad (3.288)$$

The equivalence can be shown by integrating $|\xi(t)\rangle = e^{\mathcal{H}t/i\hbar}|\mu\rangle$ and utilizing the spectral representation in Eq. (3.281). Both Eqs. (3.285) and (3.286) lead to numerically equivalent spectra. However, the latter approach requires much less computational resources. Only one matrix-vector multiplication is required per time step, with $\Delta t < \hbar/\|\mathcal{H}\|$. The Fourier integral (3.286) can be truncated due to the exponential $e^{-\gamma t}$. As a consequence the required number of time steps and thus matrix-vector multiplications is governed mainly by γ. For $\gamma = 0.1$ eV the number of steps is of the order of 10^3. Thus the scaling is of the order of $\mathcal{O}(N^2)$, compared to $\mathcal{O}(N^3)$ for the matrix diagonalization. Moreover, the matrix-vector multiplications are easily parallelized or vectorized on massively parallel or vector architectures. Hence it is possible to incorporate all relevant many-body aspects of optical excitations into *first principles* calculations.

2.3.3e Implications on the bandstructure and dielectric tensor

The impact of many-body corrections as presented above will now be discussed at the example of the lithium niobate band gap (see the present author's Publ. [12]). LiNbO$_3$ crystallizes in a trigonal structure with 10 atoms per unit cell (cf. Fig 3.11). The ground state is ferroelectric featuring the space group $R3c$. Its electro-optic, photorefractive and nonlinear optical properties are extensively used in a number of devices, such as surface acoustic wave filters and optical modulators for the telecommunications market, i. e. in mobile phones.

Given the vast range of applications, the knowledge about its electronic and optical properties is surprisingly limited. Especially the direct band gap of 3.78 eV for the ferroelectric phase – frequently cited in the literature – is actually concluded from optical experiments [169]. Thus it is affected by *electron-hole attraction effects* which may reduce the actual band gap, i. e., the difference between ionization energy and electron affinity, substantially. The issue is complicated further by the fact that actually various band gap values have been reported, all concluded from optical absorption

2.3. Optical and spectroscopic properties

experiments. They range from the indirect gap of 3.28 eV [171] to values of 4.0 or 4.3 eV [172, 173].

The theoretical understanding is also limited, since only few studies are available that address the optical and electronic properties. Most are based on a single-particle picture and neglect *quasi-particle effects* that typically widen the band gap by a large fraction of its value. The reported values range between 3.48 to 3.50 eV [174, 175]. Thus the seemingly good agreement between measured and calculated band gaps for LiNbO$_3$ may result from a fortuitious error cancellation between the possibly large *exciton binding energy* and the *electronic self-energy*. The Green's function techniques presented above are suitable to clarify this issue.

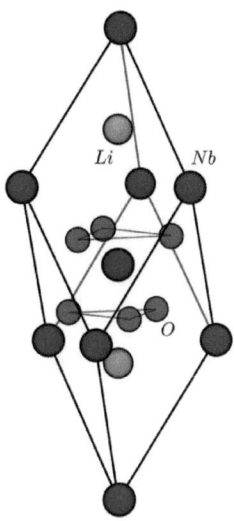

Figure 3.11: *Primitive unit cell of hexagonal LiNbO$_3$ in ferroelectric phase.*

First-principles projector augmented wave (PAW) calculations were employed using the VASP implementation of the DFT-GGA [104]. A (4 × 4 × 4) *k*-point mesh was used to sample the Brillouin zone. The electron wave functions were expanded into plane-waves up to an energy cutoff of 400 eV, with the mean-field effects of exchange and correlation in GGA modeled using the PW91 functional. First self-energy and subsequently electron-hole attraction effects were included utilizing the Green's function schemes described in the previous sections. It should be noted that the employed model dielectric function yields results that are typically accurate within about 10% of the complete calculations [163]. However, it offers another degree of freedom: by varying ϵ_∞ in Eq. (3.271) one can also account for lattice polarization effects of the system.

On this basis the Kohn-Sham energies were plotted along high symmetry lines of the hexagonal Brillouin zone (cf. Fig 3.12a, dashed blue lines). The notation of the corresponding high symmetry points is illustrated in Fig. 3.13. Within the single-particle approximation a band gap of 3.48 eV is obtained in good agreement with previous DFT calculations [174]. The valence-band maximum (VBM) occurs at the Γ point corresponding to O 2*p* states, while the conduction-band minimum (CBM) is located at $0.4 \cdot \overline{\Gamma K}$ corresponding to Nb 4*d* states.

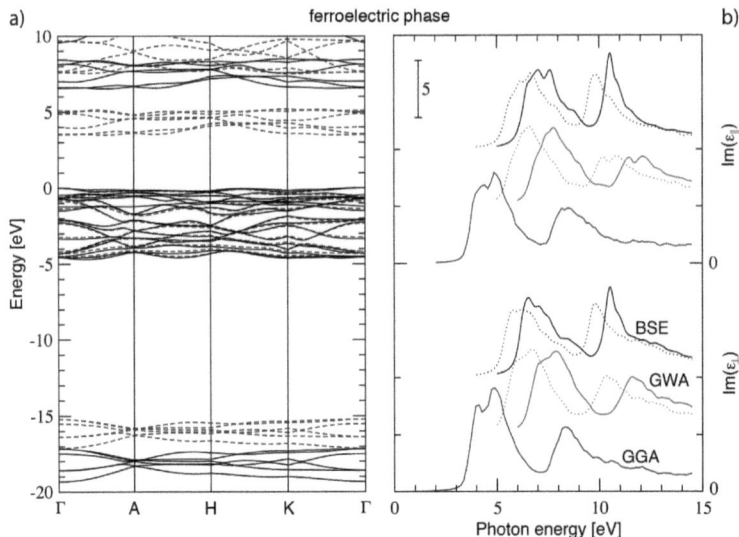

Figure 3.12: *a)* DFT *(dashed blue line) and GW (continuous black line) bandstructures of ferroelectric LiNbO$_3$ (cf. Fig. 3.13 regarding the notation of high symmetry points). Quasi-particle corrections widen the indirect band gap from 3.48 eV (DFT-GGA) to 6.53 eV (GWA). b) Parallel and perpendicular components of the dielectric function at GGA, GWA and BSE levels of theory, respectively. Solid and dotted lines indicate the results if pure electronic and electronic screening plus lattice polarizability, respectively, are taken into account.*

However, the band gap calculated within DFT-GGA does not correspond to the optical gap measured experimentally, since neither electronic quasi-particle, i. e., self-energy, effects nor electron-hole attraction, i. e. excitonic effects, are included. In order to estimate the size of quasi-particle effects GW calculations were performed. The accordingly corrected energy bands are shown as solid lines in Fig. 3.12a. The band gap is opened substantially accompanied by slight dispersion changes. For example, the CBM is relocated from $0.4 \cdot \overline{\Gamma K}$ to $0.6 \cdot \overline{\Gamma K}$. The – thus still indirect – band gap amounts to 6.53 eV.

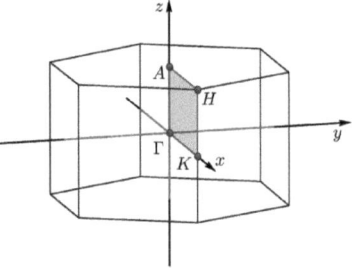

Figure 3.13: *Notation of high symmetry Brillouin zone points.*

The single-particle excitations are accompanied by a rearrangement of the remain-

2.3. Optical and spectroscopic properties

ing electrons in the solid, which screen the excited electrons and holes. However, polar materials, such as LiNbO$_3$, feature longitudinal optical phonons that give rise to macroscopic electric fields that couple to the excited electrons/holes and modify their motion. Thus it is to be expected that the lattice polarizability contributes to the dressing of the quasi-particles. This effect is not included in the GWA band structures in Fig. 3.12a, which assumes a purely electronic screening. In principle, to study the impact of the lattice polarizability upon the single-particle excitation energies the electron-phonon coupling needs to be considered. As discussed by Bechstedt et al. [177], however, the GWA in conjunction with the model dielectric function suggests a simple way to estimate the magnitude of possible lattice polarizability effects, by modifying the modeling of the screening. Rather than using ϵ_∞ for the purely electronic screening, one considers to some extent also the contribution of the lattice polarizability $\sim (\epsilon_0 - \epsilon_\infty)$. In the case of LiNbO$_3$, where the static dielectric constant $\epsilon_0 \approx 40$ is nearly one order of magnitude larger than the optical dielectric constant $\epsilon_\infty \approx 5$ [169, 170], large effects are to be expected. Assuming a partial lattice contribution to the screening, by considering a dielectric constant of 20, yields quasi-particle shifts of about half the size in comparison to those obtained for purely electronic screening. Thus the band gap assumes a value of 5.37 eV.

In the next step electron-hole attraction effects are incorporated by solving the Bethe-Salpeter equation. Fig. 3.12b shows the optical spectra according to the the three levels of theory: DFT-GGA, GWA, and BSE. Again it is not clear to which extent lattice polarizability effects contribute on the time scale of the formation of Coulomb correlated electron-hole pairs. In order to provide an estimation of their size a partial lattice contribution was assumed for both the calculation of the quasi-particle energies and the screened electron-hole attraction entering the two-particle Hamiltonian (3.279), shown as dotted lines in Fig. 3.12b. The corresponding results for a purely electronic screening are indicated as solid lines.

The spectrum obtained within DFT-GGA agrees roughly with earlier independent particle results [175, 176]. There are two main features of the optical absorption centered at about 5 and 8 eV. They arise from transitions between O $2p$ and Nb $4d$ states. The inclusion of the many-body electron-electron interaction in GWA yields a nearly rigid blue shift of the spectra by about 1.5–3 eV, depending on the screening. The electron-hole Coulomb correlation, accounted for by solving the BSE, changes the lineshape somewhat. The first peak of the low energy main feature becomes more pronounced, and the hole feature is redshifted compared to the GWA spectrum. It is now positioned at about 5.5 or 6.5 eV, depending on whether or not a partial lattice contribution is accounted for. The oscillator strength of the originally rather broad ($\sim$ 2 eV) high-energy main feature is redshifted and transformed into a single sharp peak at about 9.5 or 10.5 eV, respectively.

Compared to the experimental data for ferroelectric $LiNbO_3$, where absorption peaks at 5.3–6 and 9.2–10eV are observed [178, 179], the inclusion of self-energy and excitonic effects improves the theoretical description substantially. This concerns both the peak positions and the line shapes, which sharpen due to the inclusion of excitonic effects. Interestingly, the experimental observation that the first absorption peak is broader for $\epsilon_\parallel$ than for $\epsilon_\perp$ in ferroelectric material [179] is reproduced in the calculations that account for electron-hole interactions much more clearly than on the single-particle level of theory. The surprisingly good agreement between the Kohn-Sham gap with experiment stated often in the literature is simply fortuitous. Quasi-particle effects widen the band gap drastically beyond its DFT value. On the other hand, strong excitonic effects with exciton binding energies of the order of 1 eV can be expected.

2.4 Program packages (VASP, PWScf/WanT, DP)

The DFT calculations presented in this work are performed with the Vienna ab-initio simulation package (VASP). VASP is commercial software and can be obtained from the VASP-group at the university of Vienna [104]. It is an implementation of the plane-wave basis set scheme in reciprocal space and employs the pseudopotential approach (NC, USPP and PAW). The Kohn-Sham equations are solved for given error margins using an iterative approach (RMM-DIIS). Ionic relaxation is performed by minimizing the Hellmann-Feynman forces employing conjugate-gradients or quasi-Newton schemes. Dipole corrections allow to minimize interactions between neighbouring supercells. VASP supports the calculation of optical transition matrix elements in independent particle approximation. An unofficial version also allows to perform optical calculations within the GW approximation and to solve the Bethe-Salpeter equation.

Intraband contributions are not taken into account by VASP. The *dielectric properties* (DP) package was employed to obtain intraband corrections for the treatment of metallic cases [182]. These corrections were subsequently added to the spectra obtained by VASP.

2.4. Program packages (VASP, PWScf/WanT, DP)

The equilibrium Green's function scheme of the Landauer coherent transport formalism, as described in 4, is implemented in the *Wannier Transport (WanT)* program package. The software is released under the GNU General Public License as open source [132]. In principle it operates as a post-processing of the ground-state wavefunctions of any plane-wave DFT program package. It allows the use of both norm-conserving and ultrasoft pseudopotentials. The minimization of the Wannier functions is performed employing conjugate-gradients and steepest-descent schemes. For the present work it was interfaced to the *Plane-Wave Self-consistent field (PWScf)* program package [131]. This implementation of plane-wave DFT is also released as open source under the GNU general public license.

A multitude of tools for data processing and manipulation was self-implemented employing either PERL or Fortran.

The true delight is in the finding out rather than in the knowing.

– Isaac Asimov

CHAPTER 3
Transport properties of the clean Si(111)-(4×1)/(8×2)-In surface

Predicting the internal structure of materials based on *first principles* density functional theory (DFT) calculations is one of the prime examples of the achievement of modern solid state physics. However, in some cases – such as the presently examined In/Si(111)-(4×1)/(8×2) nanowire array – the results are not without ambiguity. This chapter presents several different structural models for the In/Si(111)-(4×1)/(8×2) nanowire array. It is discussed why an unambigous identification of the low temperature (LT) internal structure is not possible based on the total energy results alone. Subsequently, the electronic and transport properties are derived for the competing structural models and are compared to the available experimental data.

3.1 Direct approaches to structure and why they fail

Before any of the methods described in the previous chapter can be applied to predict the unknown properties of any system their accuracy has to be ensured and sensible numerical parameters need to be obtained. This is usually performed by "predicting" the experimentally well-known bulk-properties of the system of interest's constituents, in this case In and Si. At equilibrium the free energy $F(T, V)$ is minimized and for $T \to 0$ K the entropy contribution can be neglected:

$$0 \equiv p = -\left(\frac{\partial F(T, V)}{\partial V}\right)_T\bigg|_{V=V_0} \approx -\frac{\partial E_{tot}(V)}{\partial V}\bigg|_{V=V_0} \tag{1.1}$$

Thus the equilibrium condition may be reduced to the total energy E_{tot} obtained by DFT. Varying the lattice constant a of bulk-In or bulk-Si, respectively, leads to different

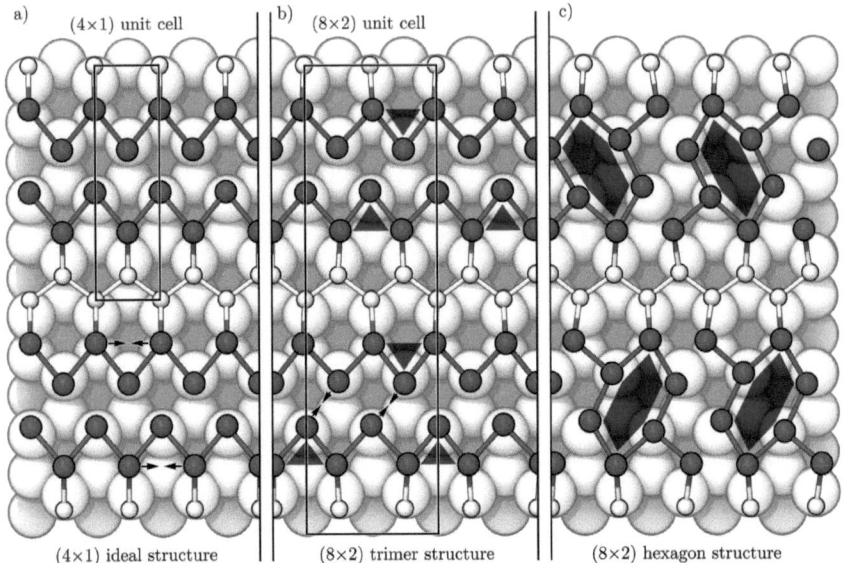

Figure 1.1: *Schematic top views of a) the ideal (4×1) model of the room temperature structure and b/c) the (8×2) trimer and hexamer structural models, respectively. Arrows indicate the movement of In atoms leading to the formation of trimers and hexamers.*

total energies that can be fitted employing the Murnaghan equation of state:

$$F(T,V) = F(T,V_0) + \frac{B_0 V}{B_0'(B_0' - 1)} \left[B_0' \left(1 - \frac{V_0}{V}\right) + \left(\frac{V_0}{V}\right)^{B_0'} - 1 \right] \quad (1.2)$$

Hence the fit yields the lattice constant a_0 and bulk-modulus B_0 at equilibrium. This approach allows to obtain a set of numerical parameters that lead to converged results that are – depending on the employed pseudopotentials – in good agreement with experiment. The present work uses PAW pseudopotentials in LDA and GGA. For these class of pseudopotentials a cutoff energy of $E_C = 250$ eV was sufficient to obtain converged results and is thus employed throughout the present work. The In/Si(111)-(4×1), (4×2) and (8×2) surfaces are simulated by repeated assymmetric slabs with six layers of Si and a vacuum region equivalent in length. Hydrogen is used to saturate the dangling bonds at the bottom sides of the slabs. The k-space integrations are performed using uniform meshes equivalent to 32 points in the Brillouin zone of the (4×1) surface.

3.1. Direct approaches to structure and why they fail

Reconstruction	GGA		LDA	
	Core d	Valence d	Core d	Valence d
(4×2) trimer	0	5	1	2
(8×2) trimer	-0.4	-5	-0.5	-0.6
(4×2) hexagon	36	48	5	15
(8×2) hexagon	25	27	-12	2

Table 1.1: *Formation energies (in meV per (4×1) unit cell) of In/Si(111) surface reconstructions relative to the ideal (4×1) chain in dependence on the treatment of the electron exchange-correlation and the explicit inclusion of the In 4d states (cf. Publ. [14]).*

Fig. 1.1a shows the structure of one In nanowire that is comprised of two ideal zigzag chains in the (4×1) unit cell parallel to the [1$\bar{1}$0] direction. This structure represents a local minimum of the surface energy. The displacement of the outer In atoms by about 0.2 Å towards each other – as indicated by horizontal arrows in Fig. 1.1a – to form pairs and finally trimers with one of the inner In atoms gives rise to another, nearly degenerate local minimum of the surface energy (cf. Fig 1.1b). The shear movement of the inner In atoms required to form hexagons – as indicated by diagonal arrows in Fig. 1.1b – is hindered by an energy barrier. After enforcing this displacement, a new energy minimum is reached, where the distance of two inner In atoms is reduced from 3.12 to 2.96 Å (cf. Fig. 1.1c). The antiphase arrangement of hexagons or trimers in adjacent wires leads to the doubling of the unit cell perpendicular to the wires.

When discussing the calculated energies for these structures, a word of caution is appropriate. The peculiar low-symmetry ground state of bulk In is related to subtle electronic effects and – due to its low stabilization energy of only 2 meV per atom – is easily distorted [183, 184]. The correct simulation of the pressure-induced phase transition of bulk In requires the inclusion of the relativistic mass velocity and Darwin terms as well as the treatment of the In 4d states as valence electrons [185]. While the inclusion of relativistic effects is out of reach for these large surface structures, the influence of the d-electron treatment (core or valence) and the XC approximation (LDA or GGA) has been probed. Similar to bulk In, the outcome of the calculation for substrate-supported In nanowires depends sensitively on the methodology used to describe the electron-electron interactions (cf. Tab. 1.1).

Within GGA, the formation of trimers rather than hexagons in (8×2) symmetry is the most energetically favored structure. This holds also in LDA, provided the In 4d electrons are explicitly included. However, once the In 4d electrons are frozen into the core within LDA the (8×2) hexagon structure becomes the most stable one by far. In contrast, the (4×2) hexagon structure is not stable in any of the examined approximations. Irrespective of the computational details, the antiphase arrangement of trimers or hexagons in (8×2) unit cells is favored over the corresponding arrangement in (4×2) symmetry. The formation of trimers or hexagons leads to local electron accumulation and loss. The alternating arrangement of these charge oscillations in neighbouring chains lowers the surface Madelung energy. Still, the overall energy gain is very small, at most 12 meV per (4×1) unit cell according to the calculations. This explains the sensitivity of the (4×1)→(8×2) phase transition with respect to external perturbations found experimentally. Tiny amounts of impurity atoms, for example, may prevent the phase transition or even revert the LT (8×2) phase to the (4×1) surface [44, 53, 64]. Such a behaviour seems hard to explain assuming the (4×1) structure to be a dynamic fluctuation between degenerate ground states with lower symmetry, as suggested in Ref. [35]. Obviously, the present findings concerning the surface energy are less conclusive than the results presented by González et al. [34, 35]. A recent study by Cho et al. also found the hexagon structure to be unstable within plane-wave DFT-GGA.

In conclusion, present day *first principles* calculations cannot describe the energetics of this system accurately enough to determine this system's geometric structure directly and unambiguously. Today the required relativistic treatment is still numerically unfeasible. These results were published in Publ. [14]. However, this systems's electronic and transport properties are highly interesting as well. Fortunately, these are rather insensitive with respect to computational details. Due to their strong dependence on the geometry, transport and electronic properties are also expected to give new insights with respect to the surface structure.

3.2 Electronic properties

To allow for a quantitative comparison of surface and bulk band structures as well as transmittances from transport calculations it is necessary to gauge the respective properties along the energy axis. A natural choice for the gauge is the effective single-particle potential $v_{eff}(r)$ of bulk-Si. However, before the actual gauging the Si-bulk band structure needs to be projected onto the surface Brillouin zone of the respective structural model for the In/Si(111) surface. After filling the unit cell of the respective structural model completely with bulk Si the band structure is calculated along the surface Brillouin zone for different fixed vectors $k_\perp$ perpendicular to the projection

3.2. Electronic properties

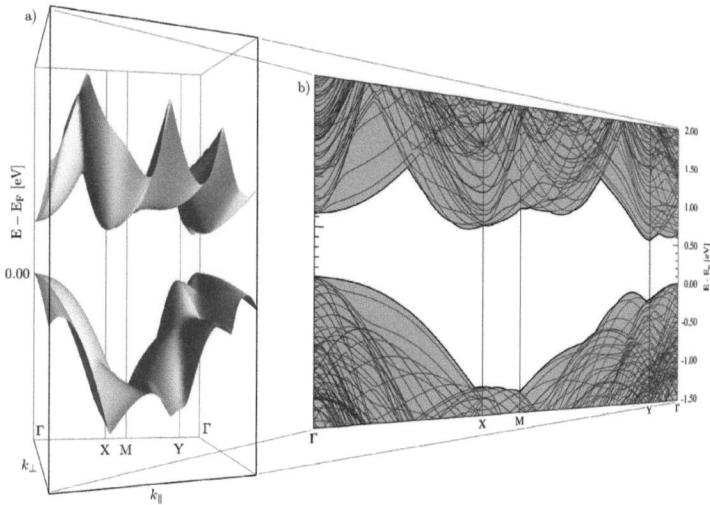

Figure 2.2: *Projection of the Si-bulk band structure onto the surface Brillouin zone of the Si(111)-(4×1)-In structural model. The projection is performed along the $k_\perp$-direction.*

plane. Fig. 2.2a shows the result for the (4×1) unit cell. The actual projection is subsequently performed along $k_\perp$ (cf. Fig. 2.2b).

Averaging the Kohn-Sham effective single-particle potential in plane with the In nanowires for both the (4×1) surface and bulk-Si in the equivalent unit cell yields the sinusoidal oscillations shown in Fig. 2.3a. Each one of the deep minima corresponds to the location of one Si bilayer along the surface's normal axis. The more shallow one of the (4×1) surface (indicated in red) represents the In nanowires themselves. The actual gauge can then be performed by shifting the In/Si(111)-(4×1) band structure along the energy axis by the difference ΔE of the Si bilayer's effective single-particle potentials of both systems. This procedure yields the (4×1) surface band structure shown in Fig. 2.3b, with the projected Si-bulk states indicated by the filled gray area. The valence band maximum is assumed as the energy zero. Applying the analogous procedure to the (4×2)/(8×2) structural models leads to the surface band structures shown in Fig. 2.4a-d.

The (4×1) structural model exhibits three quasi-1D In surface bands (indicated in blue) with strong dispersion along the chain direction (ΓM), but weak dispersion perpendicular to the chains (XM), in agreement with earlier work [29]. Upon trimer-

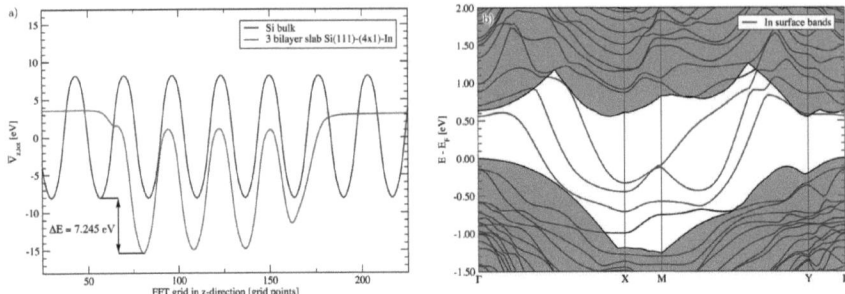

Figure 2.3: *a) Total effective single particle potential averaged in plane with the nanowires for both the Si(111)-(4×1)-In surface and Si-bulk in the equivalent unit cell. b) Band structure of the (4×1) structural model of the room temperature phase. Regauging of the In surface bands with respect to the Si bulk band structure was performed as indicated in Fig. 2.3a.*

ization (cf. Fig. 2.4a/b) local gaps are opened near the X and M points. However, one surface band still crosses the Fermi level, in accordance with Ref. [186]. Hence the trimer structures do not exhibit a fundamental energy gap in clear contradiction to the experimental findings in Refs. [30, 31]. The band structure calculated for the hexagon model in (4×2) symmetry largely reproduces the results by González et al. [35]. It should be noted, however, that González et al. in contrast obtained the valence band maximum (VBM) at the Γ-point, while in the present work the VBM is located near the X-point. This numerical artifact is assumed to be caused by the incompleteness of atomic orbitals as a basis set, which are employed in the work by González et al. While the hexagon model was originally discovered within this local orbital basis set, the structural relaxations may not be entirely accurate due to such inconsistencies in the electronic structure. The plane-wave DFT results in the present work can be expected to be more reliable.

Remarkably, the small changes of the chain geometry and the interwire interaction upon arranging the hexagons in an antiphase (8×2) symmetry lead to rather distinct changes in the band structure, in addition to the band folding. The gap between occupied and unoccupied surface states near Γ is widened, and the fundamental gap near X′ shrinks and becomes more direct. Given the underestimation of the band gap within DFT, the calculated band gap of 0.05 eV (GGA, $4d$ in valence) is consistent with the experimental findings of 0.16 eV from STM [42] and 0.3 eV from surface conductivity measurements [44]. The position of the calculated maximum of occupied bands near X′ is confirmed by photoelectron spectroscopy [30, 36].

3.3. Transport properties

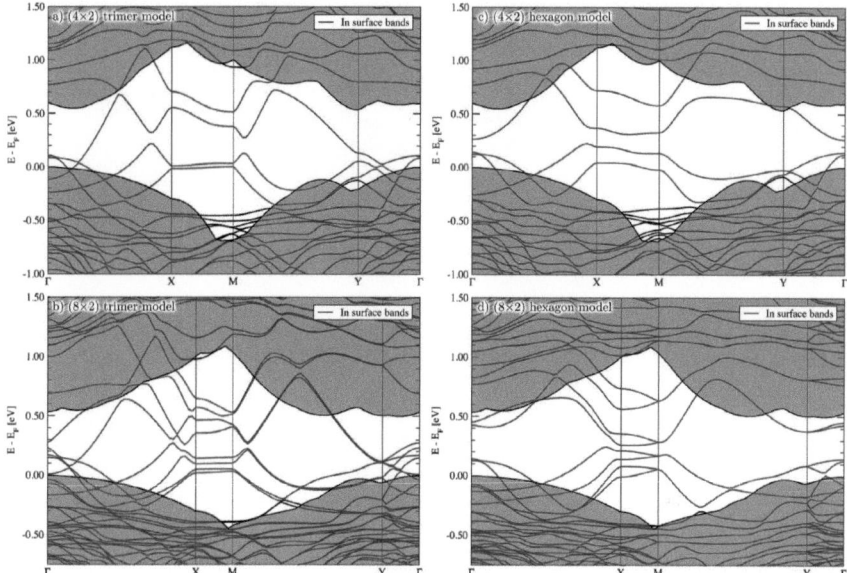

Figure 2.4: *Band structures of the a/c) (4×2) and b/d) (8×2) trimer and hexagon models, respectively. Regauging of the In surface bands with respect to the Si bulk band structure was performed as indicated in Fig. 2.3a.*

3.3 Transport properties

Based on the derivation of the electronic properties as discussed in the previous section, one can now calculate the electron transport properties for these systems. However, performing these transport calculations is not as straight forward as in the case of DFT calculations. Therefore a brief discussion of some numerical modeling aspects is in order before proceeding to the actual results.

3.3.1 Computational details

As described in chapter 2.2.4 electron transport is conveniently and naturally expressed in terms of a localized basis set. Since the plane-wave basis of DFT is a priori delocalized, it is necessary to change the basis set. In this case Wannier functions (WFs) were chosen because they can be obtained from plane-waves by a simple unitary transformation and span exactly the same Hilbert space. Thus there is no approximation and no loss of the *ab initio* accuracy involved in this basis set transformation. However, the resulting Wannier functions are non-unique due to the *k*-

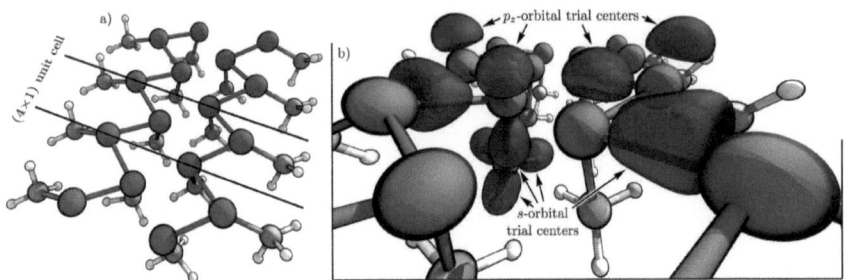

Figure 3.5: *a) Model system for the unreconstructed (4×1) surface with nearest neighbour Si atoms only. The Si dangling bonds are saturated by hydrogen. b) Orbital symmetries of Wannier trial centers and isosurfaces of the resulting Wannier functions after the minimization procedure.*

dependent arbitrary phase factor of the Bloch functions and the invariancy of the electronic structure problem as opposed to the transformation $\psi_{nk} \to e^{\phi_n(k)}\psi_{nk}$ (cf. chapter 2.2.4). Thus the requirement of a *localized* basis set implies that a basis set of *maximally localized Wannier functions* must be identified within the infinite number of possible WF basis sets.

Due to the size and complexity of this minimization problem's phase space a fully automatic solution is impossible in all but the most simple cases. After choosing an energy window within the electronic bands are to be described by WFs, the localization procedure must be supplied with *trial Wannier functions* – one for each electronic band – whose symmetries and positions should be sufficiently close to the ones of the minimized, optimal Wannier functions. In this case atomic orbitals were employed as Wannier trial centers. The exact choice of these trial centers is highly critical with respect to whether the WFs converge towards an acceptable solution or not. Beginning with a very small system – i. e. a single nanowire in a (4×1) unit cell with the nearest neighbour Si atoms only (cf. Fig. 3.5a) – one can conveniently test different trial centers for their convergence properties. After much experimentation sensible trial centers for the In nanowires turned out to be s-orbitals placed on each of the bonds and p_z-orbitals oriented along the surface normal located directly at the In atoms. This way convergence can be achieved within a reasonable amount of time while all relevant bands are represented. It is to be noted that the In $4d$ bands are treated as frozen into the core since they do not contribute to the conductance. Thus no d-orbital trial centers are required.

If the system under examination contains extended vacuum regions – as it is the case

3.3. Transport properties

for the In nanowires – individual WFs often wander off into these vacuum regions. By delocalizing completely they significantly improve the localization properties of the remaining WFs. However, from the employed tight-binding point of view the resulting WF set is completely useless, of course. To suppress this behaviour the WFs need to be weighted by penalty functionals that effectively attach a spring potential to the WF centers (cf. section 2.2.4c). The amplitude of this penalty functional has to be chosen with care. A too small amplitude will not prevent the WF from wandering off, while a too large amplitude prohibits the localization procedure from ever attaining convergence. To ensure as accurate results as possible the states contained within an interval of ± 1.5 eV around the Fermi energy were treated as frozen during the disentangle-procedure (cf. section 2.2.4c). Since most of the bands within this interval are fairly well separated with very few degeneracies the WFs localization properties are barely affected.

The resulting Wannier functions after the minimization procedure are depicted in Fig. 3.5b, with the average spread amounting to 5.5 Å. Since the tight-binding formalism requires any basis function overlaps to be negligible beyond nearest neighbour unit cells this calculation is still slightly out of convergence with respect to the unit cell size (5.5 Å average spread as opposed to the unit cell's shortest lenght of 3.86 Å). While this treatment is adequate for convergence studies, the later calculations need to be performed in larger unit cells.

It should be stressed again that this treatment is formally exact. However, the condition that the resulting Wannier function set has to fulfill the tight-binding condition of negligible interactions between next-nearest neighbour cells and the multitude of numerical parameters make the derivation of a suitable WF set indeed extremely error-prone. Thus the final set of WFs needs to be checked very thoroughly. The WFs must describe exactly the same system within the tight-binding approach as DFT does within its associated plane-wave basis. Hence a very useful test is to recalculate the system's electronic properties from the Hamiltonian in the Wannier basis. Naturally, the results must be identical to the original DFT data. Both the electronic structures obtained from DFT (circles) and the WF Hamiltonian (solid lines) for this model system (cf. Fig. 3.5) are shown in Fig. 3.6. Around the Fermi energy they match very well but increasing deviations occur towards larger energies. This is due to the too small size of the unit cell as well as because the states are not treated as frozen anymore above 1.5 eV.

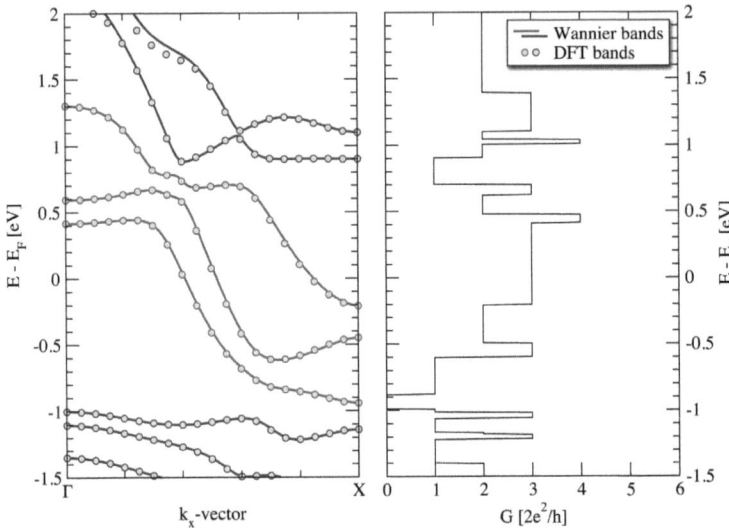

Figure 3.6: *Comparison of band structures obtained from DFT (circles) and within the Wannier function basis (solid lines), respectively, and conductance G(E) of the (4×1) RT phase with nearest neighbour Si atoms only (cf. Fig. 3.5). The three In surface bands are marked in* red.

In this simple test case the conductor and the leads are assumed to be identical. Thus there is no contact resistance and the respective system's conductance $G(E)$ must be identical to the number of modes – in this case the number of bands – at any given energy E. Comparing the electronic structure with the conductance obtained from the WF basis set Green's function approach in Fig. 3.6 shows that this condition is satisfied. Thus the description of the system within the WF basis set is indeed accurate.

So far the (4×1) RT model system contained only the nearest neighbour Si atoms as substrate. Fig. 3.7 shows the conductance spectra for the same model system containing up to 2 bilayers of Si substrate. The first spectrum is identical to the right hand side of Fig. 3.6. It is obvious that the In surface states depend very little on the substrate. Within an interval of ±0.2 eV around the Fermi energy the obtained conductances are identical. Thus for the In/Si(111)-(4×1) surface electron transport is indeed well described by model structures that contain only the In and nearest neighbour Si atoms with the remaining Si dangling bonds terminated by hydrogen.

One approximation that was introduced with this model system is its strict quasi one-dimensionality, as it does not represent a surface anymore. Within the employed

3.3. Transport properties

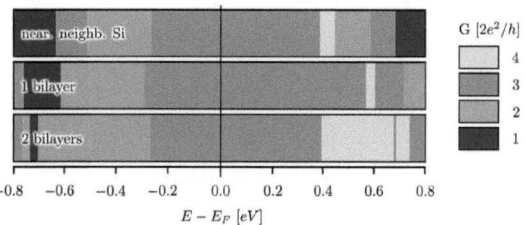

Figure 3.7: *Quantum conductance spectra for electron transport along the chain direction calculated for In/Si(111)-(4×1) model structures containing only the nearest Si neighbour atoms or 1 and 2 substrate bilayers, respectively (cf. Publ. [13]).*

Wannier function methodology a two-dimensional system is treated as an inifinite number of independent parallel one-dimensional systems. In essence a huge number of k-points perpendicular to the transport direction would be required for an accurate treatment. For the structures of interest in the present work such a treatment is numerically unfeasible. A better approach to retain at least some of the original surface reconstruction's interchain coupling is to perform all calculations within the (8×2) unit cell, irrespective of the actual translational symmetry. This treatment also ensures that the unit cell is large enough with respect to the tight-binding condition. On this basis it is now possible to perform electron transport calculations for the different In nanowire array's structural models.

3.3.2 Conductance spectra and discussion

Fig. 3.8 shows the calculated quantum conductances of five infinite nanowires that model the In/Si(111) surface. The atomic coordinates of the In and neighbouring Si atoms were taken from the relaxed surface structures (GGA, In 4d electrons in valence). The electron transport calculations were performed as described in the previous section, employing a 20×1×1 k-point grid. Obviously, the energy dependent transmittance is very sensitive with respect to the wire geometry, as expected from the band structures discussed in section 3.2. Because of the free electrons in the metallic In wires of the In/Si(111)-(4×1) surface, its transmittance is nearly constant over the energy range of the Si band gap. The value obtained in this calculation somewhat overestimates – by roughly a factor of 2 – the experimentally determined surface-state conductance in the RT regime [33, 43, 44]. This is partially an effect of the metallic contacts and their scattering as well as related to the thermal dissipative scattering to phonons at finite temperature.

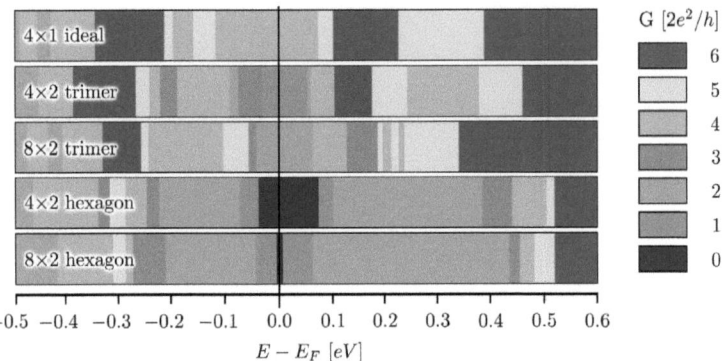

Figure 3.8: *Quantum conductance spectra for electron transport along the chain direction calculated for In/Si(111) model structures (cf. Publ. [14]).*

The formation of In trimers with (4×2) or (8×2) symmetry does not quench the conductance but leads to a reduction for energies very close to E_F, i. e. within about 10^{-1} eV. For energies farther below or above E_F, the transmittance resembles those of the ideal chains. Only some conduction channels shift slightly in energy or are somewhat reduced in magnitude. The conductance changes upon hexagon formation are significantly more pronounced. In both the (4×2) and the (8×2) symmetry, the transmittance through surface states is strongly reduced over an energy range of nearly 1 eV. For energies of 10^{-1} and 10^{-2} eV around E_F there is zero conductance for hexagons arranged in (4×2) and (8×2) symmetry, respectively. These findings may well account for the vanishing surface-state conductance measured for the LT phase [33, 44] of the In nanowires (cf. Publ. [14]).

The above work also represents a solid foundation for determining the impact of doping on the wires' structure and conductance, as examined in the following chapter.

Electricity, n.: The cause of all natural phenomena not known to be caused by something else.

– Ambrose Bierce

CHAPTER 4
Transport properties of the doped Si(111)-(4×1)/(8×2)-In surface

This chapter begins with the determination of the preferred adsorption positions and geometric structures for H, O, In and Pb impurity atoms. The influence of these defects on the In nanowire conductance is evaluated employing a lead-conductor-lead partitioning of the system, where the In chain segment with the defect forms the conductor and the semi-infinite leads are modeled with ideal In nanowires. Subsequently, several different conductance affecting mechanisms are presented and analyzed in detail.

4.1 Adsorption of impurity atoms on In/Si(111)-(4×1)

As discussed in section 3.1 DFT calculations have difficulties to describe accurately the subtle energetics of the temperature-induced (4×1)→(8×2) phase transition (cf. Ref. [35], Publ. [14]). On the other hand, all measured properties of the RT (4×1) phase are well reproduced within DFT [52, 187, 188, 189]. Therefore, and because the potential energy surfaces (PES) of the adatoms considered here (cf. Figs. 1.2 and 1.1) turn out to be highly corrugated, the following DFT calculations for ideal and adatom-perturbed In/Si(111)-(4×1) surfaces are expected to be accurate.

4.1.1 Potential energy surfaces (PES)

As a starting point the In/Si(111)-(4×1) surface needs to be probed individually for each impurity atom species for energetically favourable adsorption positions. By calculating the surface adsorption energies at fixed lateral positions one can construct a "map" of the adsorption energetics, the so-called *potential energy surface* (PES). While the adatom itself was fixed laterally, it was free to relax along the surface's normal

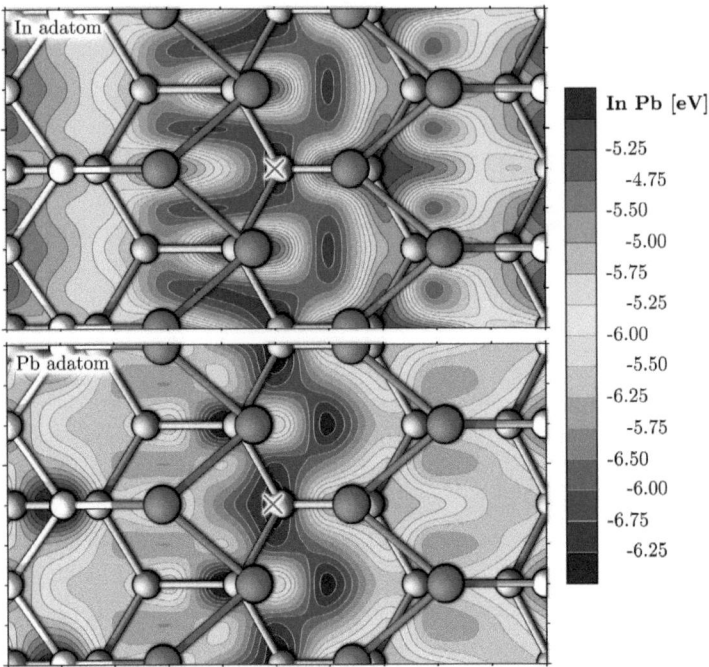

Figure 1.1: *Potential energy surfaces calculated for indium and lead adatoms on the In/Si(111)-(4×1) surface The minimum energy positions are marked by × (cf. Publ. [11]).*

direction. No restrictions were applied to the surface's degrees of freedom, with the exception of the lowest Si layer fixed at Si-bulk coordinates. These calculations were performed in a (4×3) unit cell with 3 bilayers of Si. A 8×2×1 k-point mesh was employed to sample the Brillouin zone at an energy cutoff of 400 eV for the wavefunction expansion. The In 4d states were treated as core electrons. The adsorption energy E_{ads} is given by

$$E_{ads} = E_{InSi(4\times1)+adatom} - E_{InSi(4\times1)} - E_{adatom} \qquad (1.1)$$

where $E_{InSi(4\times1)}$ and $E_{InSi(4\times1)+adatom}$ denote the total energies of the clean and impurity atom adsorbed In/Si(111)-(4×1) surfaces, respectively. E_{adatom} is the total energy of the single isolated adatom.

Figs. 1.1 and 1.2 show the calculated energy landscapes in terms of E_{ads}. In and Pb prefer a position between neighbouring In chains. In case of Pb an adsorption is also

4.1. Adsorption of impurity atoms on In/Si(111)-(4×1)

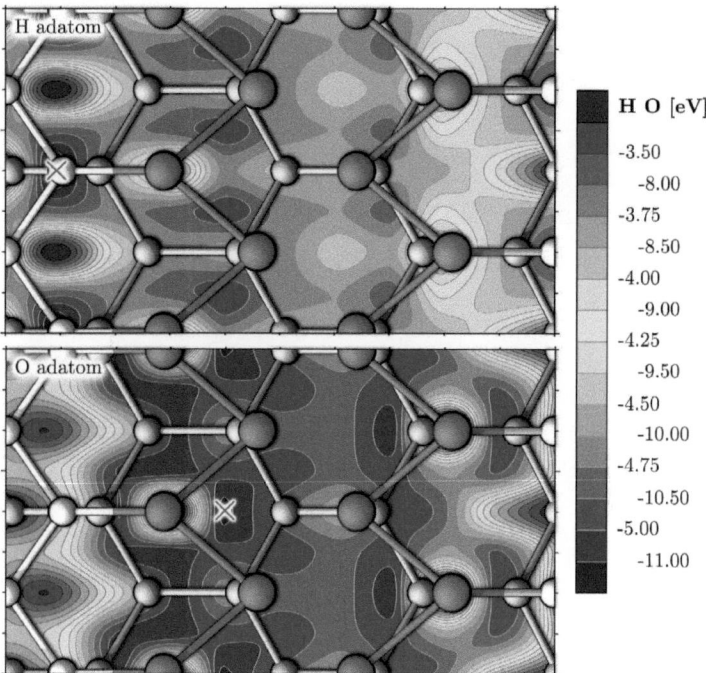

Figure 1.2: *Potential energy surfaces calculated for hydrogen and oxygen adatoms on the In/Si(111)-(4×1) surface. The minimum energy positions are marked by × (cf. Publ. [11]).*

possible besides the In chain next to one of the inner In chain atoms. However, this minimum on the PES is slightly shallower. O adsorbs threefold coordinated on top of one of the In chains. Similar to Pb, O also adsorbs between neighbouring Si and In chains. For H, however, the adsorption on the In chains represents only a local minimum on the PES. Instead H prefers to bond directly to one surface Si atom.

In Ref. [64] a possible mobility of O adatoms on the In/Si(111)-(4×1) surface was suggested as one approach to explain the experimentally observed increase of the critical temperature T_C. In contrast to the other adatom species oxygen is strongly attracted along the entire range of the In chains. However, a closer look reveals rather high diffusion barriers on the order of 0.5 eV between the preferred adsorption sites. Thus oxygen is expected to be immobile on the In/Si(111)-(4×1) surface. Diffusion barriers of similar magnitude exist for the other adatom species as well (H: 0.5 eV, In: 0.25 eV, Pb: 0.25 eV), effectively immobilizing any of these defects at room temperature.

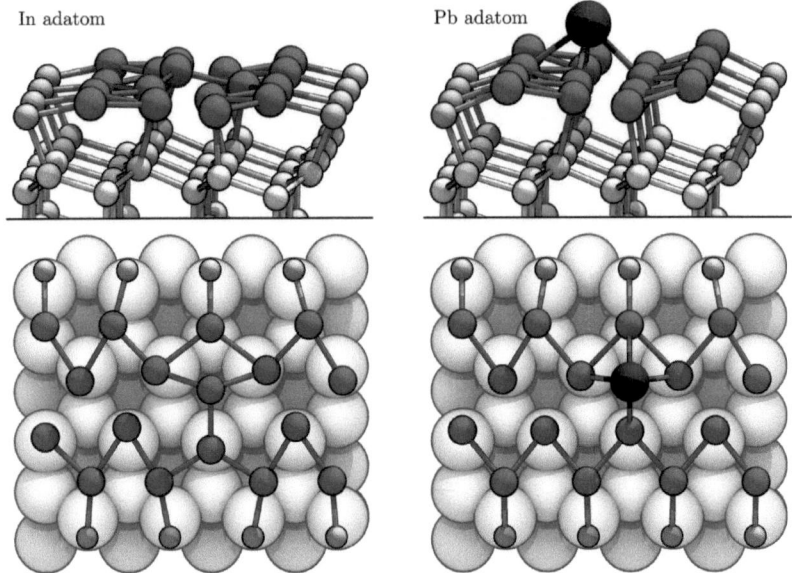

Figure 1.3: *Schematic top and side views of In and Pb adatoms adsorbed on the In/Si(111)-(4×1) surface.*

4.1.2 Structural properties

Placing the respective adatom species at the minimum energy positions marked in Figs. 1.1 & 1.2 and allowing the system to relax freely yields the geometric structures illustrated in Figs. 1.3 and 1.4. While the preferred lateral positions of O, In and Pb are similar, the adsorption geometries are not. Both the O and In adatoms are located almost in-plane with the In chain atoms. This affects the structural relaxation of the In nanowire considerably:

- In strongly displaces the adjacent In chain atoms in an outward direction.

- Pb prefers a pyramidal configuration on top of the In nanowire that induces comparatively small deformations.

- H adsorbs on one of the surface Si atoms. Thereby the Si-In bond is broken shifting the In atom slightly inwards. Otherwise the induced chain deformation is small.

- For the O adatom the reverse is the case with respect to In deposition: O strongly attracts the adjacent In chain atoms, moving them inwards.

4.1. Adsorption of impurity atoms on In/Si(111)-(4×1)

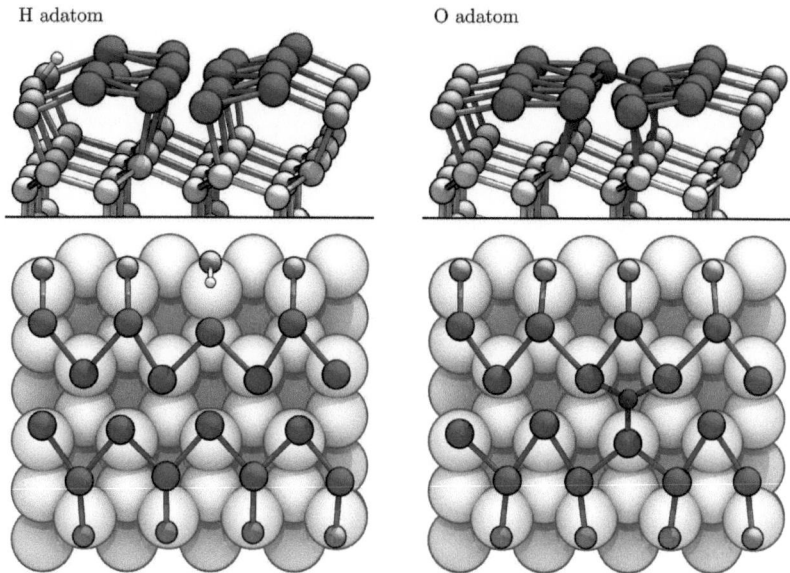

Figure 1.4: *Schematic top and side views of H and O adatoms adsorbed on the In/Si(111)-(4×1) surface.*

This is reflected in the standard deviations of the In-In bond length distribution σ and the average atomic shifts $\bar{\Delta}$ compiled in Tab. 1.1. The calculated positions of O and Pb agree well with the available experimental data in Refs. [58, 61, 64]. In case of H the actual position is somewhat more difficult to determine, since measurements usually observe the H-induced structural deformations but not the H adatom itself. Nevertheless, simulated STM images for H atoms adsorbed on the Si chains agree well with experimental STM images [59, 60, 64]. No experimental data of sufficiently high resolution could be found for In deposition.

Adatom species	$\sigma[\text{Å}]$	$\bar{\Delta}[\text{Å}]$
Ideal	0.01	0.00
H	0.04	0.21
O	0.11	0.24
In	0.12	0.38
Pb	0.07	0.16

Table 1.1: *Standard deviation σ of the In-In bond length distribution and average shift $\bar{\Delta}$ of the In chain atoms for ideal and defect-modified nanowires.*

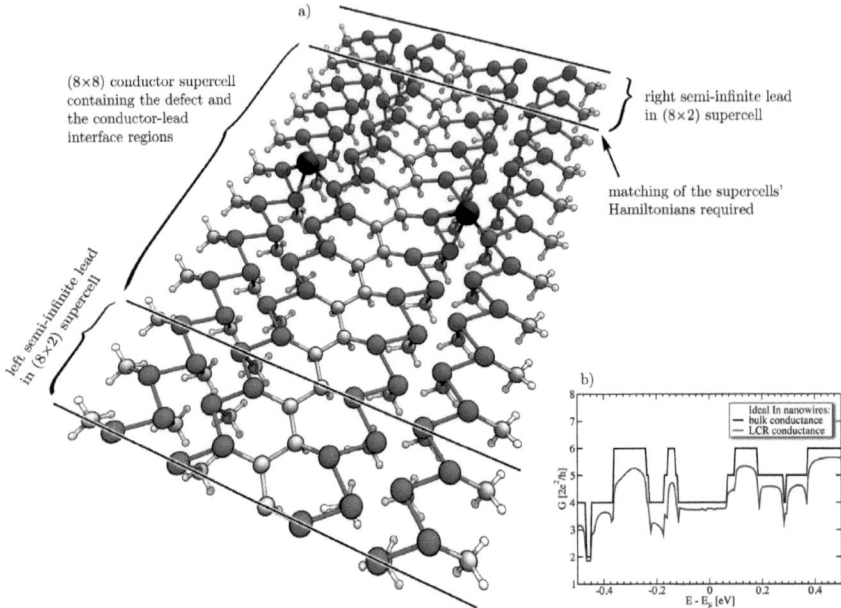

Figure 2.5: *a) Partitioning of the* left lead – conductor – right lead *(LCR) system into a (8×8) conductor cell and separate (8×2) cells for the semi-infinite leads. The conductor supercell should also contain part of the leads to treat the interface regions from first principles. b) Conductances of the ideal unperturbed In nanowires calculated for the bulk case in a (8×2) supercell (black line) and employing the full LCR formalism with (8×2)/(8×8) supercells (blue line), see text.*

4.2 Transport properties

4.2.1 Computational details

Starting from the relaxed adatom positions, the influence of the perturbation on the In nanowire conductance is calculated employing a *left lead – conductor – right lead* (LCR) partitioning of the system. Here the In chain segment with the impurity forms the conductor within a (8×8) wire segment. The semi-infinite leads are modeled with ideal In nanowires in a (8×2) wire segment (cf. Fig. 2.5a). For the leads this is exactly the same calculation as for the ideal In nanowires in the previous chapter. A $16 \times 1 \times 1$ k-point mesh and an energy cutoff of 250 eV were found to result in converged spectra.

It is to be noted that the conductor supercell should also contain parts of the leads. Thus the lead-conductor interface regions can also be treated from first principles,

4.2. Transport properties

as both the conductor itself and the interface regions are simulated within the same calculation. The amount of lead layers to be included within the conductor supercell should be sufficiently large, so that the local electronic structure at the edges of the conductor supercell resembles the electronic structure of the bulk leads. This convergence can be controlled by comparing the respective WF Hamiltonian matrix elements at the edges of both the lead and conductor supercells. A (8×8) supercell was found to be sufficient.

In the next step the Hamiltonian matrix of the whole LCR system needs to be constructed. Constructing the matrix is performed manually by extracting the matrix elements from the separate lead and conductor calculations. This process is highly susceptible to both human and numerical errors. Thus it is very useful to perform the LCR transport calculation first for a system whose conductance is already known. The ideal In nanowires are a suitable test system, as their bulk conductance was derived in the previous chapter employing a (8×2) supercell. Filling the conductor supercell with an (8×8) section of the ideal In nanowires and attaching a (8×2) supercell for the semi-infinite leads is a rather complicated way to describe the physically equivalent system. Hence the resulting conductance must again be the same.

Fig. 2.5b shows the conductances of the ideal In nanowires derived from the (8×2) bulk (black line) and (8×2)/(8×8) LCR calculations (blue line), respectively. The results show slight deviations from each other due to numerical noise and a small mismatch in energy between the Hamiltonian matrix elements of the lead and conductor calculations. As the matrix elements are expressed with respect to the Fermi energy of the respective calculation, different calculations with different Fermi energies may require a rigid shift of the matrix elements to avoid such a mismatch. However, in this case the mismatch is very small indeed and the resulting conductances match reasonably well. Especially within an interval of ±0.1 eV around the Fermi energy the agreement is very close (4.0 as opposed to 3.8 in units of $2e^2/h$). Due to the In wires' metallicity conductance measurements are always performed close to E_F. Larger voltage drops across the wires would induce currents causing enough heating of the array to ultimately destroy the wires. Thus the voltage range of about ±0.1 eV is the most interesting one and the simulation can be considered accurate.

After performing these tests the ideal In nanowires in the conductor supercell can now be replaced by the respective adatom-decorated and distorted In nanowires. The calculational formalism remains identical.

4.2.2 Conductance spectra

Employing the previously described formalism, the impact of the different adatom-induced perturbations on the In nanowire conductance is calculated. In Fig. 2.6 the resulting conductance spectra of ideal and adatom-perturbed In nanowires is illustrated. Compared to the calculations for the ideal structure, a reduction of the conductance at E_F by more than one third compared to the ideal chains is calculated for the case of In adatoms. While the experimental conditions are not sufficiently well defined to allow for a quantitative comparison, the calculated reduction is of the same order as measured [44]. Interestingly, apart from hydrogen which does not substantially modify the electron transport properties, distinct conductance drops are predicted as well for Pb and in particular for O, see Tab. 2.2. On first glance one might want to explain these findings with a reduced density of states (DOS) at the Fermi energy E_F. However, as shown in Fig. 2.6 and Tab. 2.2:

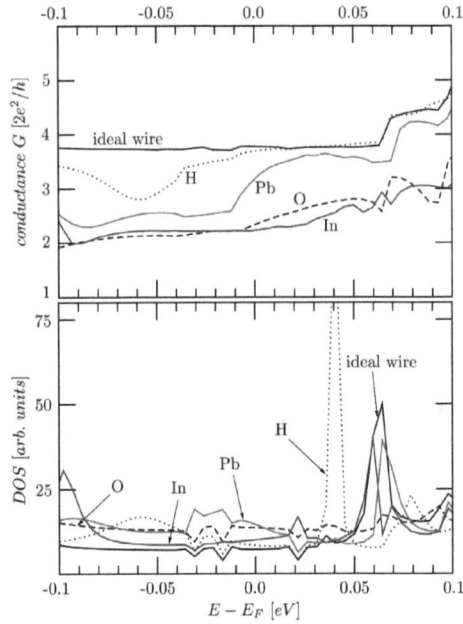

Figure 2.6: *Quantum conductance spectra for electron transport along the wire direction (upper part) and total density of states (lower part) calculated for ideal and adatom-modified In/Si(111) structures (cf. Publ. [11]).*

Adatom species	$G\ [\frac{2e^2}{h}]$	DOS [a. u.]
Ideal	3.75	0.70
H	3.60	1.44
O	2.43	1.22
In	2.31	0.95
Pb	3.11	1.23

Table 2.2: *Average quantum conductance G and DOS in the energy interval ±0.05 eV around the Fermi energy.*

4.2. Transport properties

- The DOS increases irrespective of the specific adatom deposited

- The DOS of the perturbed nanowires shows at best a very weak correlation with the conductance

For example, the DOS at E_F is nearly equal for Pb and O. However, O is far more effective in reducing the conductance. Obviously, the DOS does not suffice to understand the trend of the conductance change. From a classical point of view these findings are rather puzzling: In spite of "making the wire thicker" and thus introducing many new electronic states within the wire the conductance is significantly lower. However, before proceeding to a detailed examination of the involved quantum conductance quenching mechanisms a few remarks with respect to the comparison of experimental and theoretical conductance spectra are in order.

4.2.3 Adatom-localized phonon modes

As discussed in the previous chapter, the transmittance at E_F somewhat overestimates – by roughly a factor of 2 – the experimentally determined surface state conductance in the RT regime. Besides effects of the contacts this can be attributed to thermal dissipative scattering due to phonons at finite temperature. These effects arise as well for perturbed In nanowires. However, the phonon scattering – neglected in Fig. 2.6 – will modify the electron transport to different degrees.

In order to estimate the influence of adatom-localized surface phonons *frozen-phonon* calculations were performed for the uppermost layer of each adatom-deposited structure, respectively. According to the frozen-phonon approach each atom is displaced systematically once in each spatial direction to obtain the forces. In harmonic approximation the forces are given by:

$$F_i(u_j) = -k_{i,j}u_j + l_{i,j}u_j^2 \qquad (2.2)$$
$$F_i(-u_j) = +k_{i,j}u_j + l_{i,j}u_j^2 \qquad (2.3)$$

The linear force coefficient can be extracted from two elongations:

$$k_{i,j} = \frac{F_i(-u_j) - F_i(u_j)}{2} \qquad (2.4)$$

The atomic and cartesian indices were combined to simplify the notation. Symmetrizing the force constant matrix $\mathcal{K}$ defined by $k_{i,j}$ and employing a harmonic approach

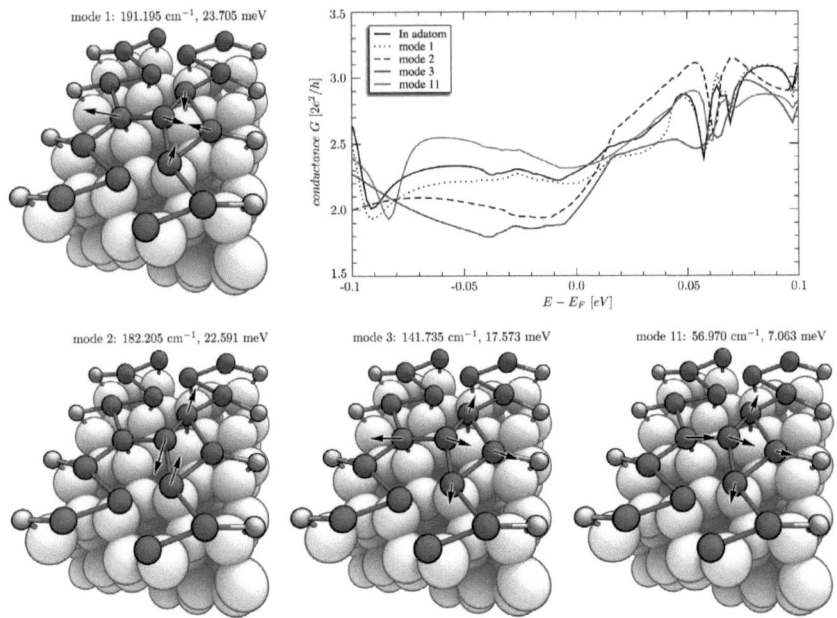

Figure 2.7: *Adatom-localized phonon modes for the In-deposited In nanowires and their impact on the quantum conductance for room temperature occupation, see text.*

$$u' = u e^{-i\omega t} \quad (2.5)$$

one can obtain the eigenmodes and -frequencies by solving the associated algebraic eigenvalue problem:

$$\omega^2 \mathcal{M} u = \mathcal{K} u \quad (2.6)$$

The resulting adatom-localized eigenvectors and -frequencies for In deposition are illustrated in Fig. 2.7. Naturally, many more modes are in existence. However, since they are not localized around the adatom they are similar for all examined structures and thus also influence electron transport to similar degrees. To estimate an upper bound for the influence of adatom-localized modes the nanowire atoms are elongated along the scaled eigenvector $b \cdot u$ so that the mode's mean potential energy $\bar{V} = \frac{1}{2}V(b \cdot u)$ is set to $\bar{V} = \frac{1}{2}k_B T$. The elongation-dependent potential energy $V(u)$ of a mode is given by:

$$V(u) = \sum_i v_i(u), \quad [v_1, v_2, ..., v_i](u) = \frac{1}{2}\omega^2 \mathcal{M} u^2 \quad (2.7)$$

4.3. Conductance quenching mechanisms 115

Subsequently electron transport calculations were performed for the accordingly distorted structures. Fig. 2.7 shows the resulting conductances for In deposition. Interestingly, there are also modes that increase the nanowire conductance. However, from the point of view of the phase transition this behaviour is understandable: At LT the nanowires are semiconducting. Thus there must be phonon modes in existence, that transform the LT structure to the metallic RT structure (these modes are identified and discussed in detail in chapter 6). Nonetheless, most of the examined adatom-localized modes decrease the conductance. Averaging over all 4 modes in an interval of ± 0.05 eV around the Fermi energy yields an average conductance drop of 9% for In deposition.

The impact is different, however, for H deposition, which exerts practically no influence at all. This can be understood in terms of the H adatoms eigenfrequencies: The hydrogen modes are much too high in energy to be excited at room temperature. Due to the large computational cost the impact of O and Pb induced modes has not been calculated so far. However, the derivation of the temperature dependent conductance for doped In nanowires might be an interesting project for the future. In chapter 6 an approach is presented to perform such calculations for the unperturbed In nanowires. In principle the same approach can be applied to perturbed In nanowires as well.

In conclusion, the impact of the respective impurity atoms on the temperature dependent conductance can be expected to be rather pronounced. In comparison with experimental data the different effects of different impurities at finite temperatures must be kept in mind. Now, the next section proceeds to a detailed examination of the involved quantum conductance quenching mechanisms.

4.3 Conductance quenching mechanisms

4.3.1 Local density of states (DOS)

While the total DOS proved to be of little help in understanding the conductance drops, one might suspect the *local* DOS to be more revealing. Fig. 3.8 shows a difference plot representing the adsorption induced changes of the local DOS at E_F projected onto the plane of the Si(111) surface. In all cases a distinct and adatom-specific LDOS modification upon adsorption is observed. I. e., In causes a sharp local DOS increase close to the adatom that is accompanied by a DOS depletion at the next nearest neighbour distance. The adsorption of O leads to a DOS redistribution from

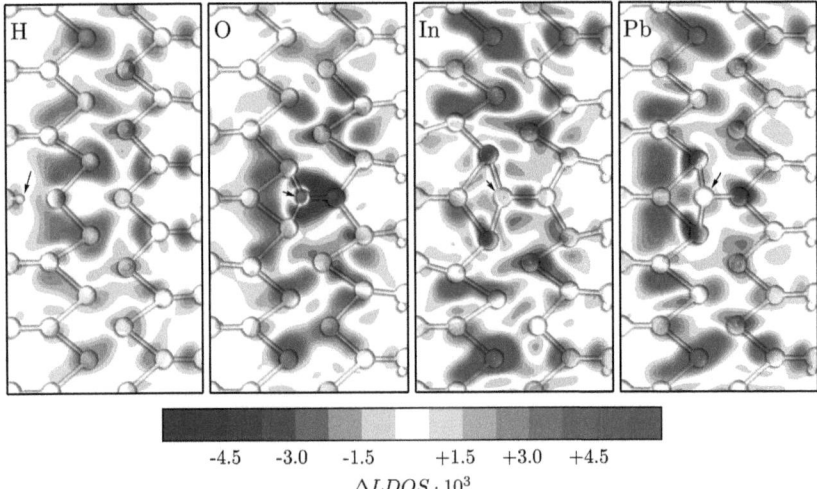

Figure 3.8: *Adsorption induced changes (in arbitrary units) of the local density of states (LDOS) at E_F projected on the plane of the Si(111) surface. Negative and positive values indicate local DOS depletion and accumulation with respect to the ideal In/Si(111) system, respectively (cf. Publ. [11]). Adatom positions are marked by arrows.*

the neighbouring to the adatom decorated In chain. Even hydrogen – that does not affect the nanowire conductance and adsorbs on one of the Si atoms rather than on the In chain – clearly affects the nanowire LDOS. From a qualitative point of view adatom-induced changes in the local DOS may seem suggestive for understanding the influence of adatoms on the nanowire conductance. However, no quantitative correlation with the conductance could be deduced.

4.3.2 Potential-well scattering

In case of CO adsorption on substrate-supported Au chains, the drastic conductance drop could be traced to the deep potential well arising at the adsorption site [190]. In order to see whether a similar mechanism acts here, the local effective single particle potential $V_{eff}(r)$ was extracted from the DFT calculations. This potential was averaged subsequently in a plane perpendicular to the nanowire direction chosen large enough to contain – within their covalent radii – the nanowire In atoms, as well as the adatoms.

4.3. Conductance quenching mechanisms

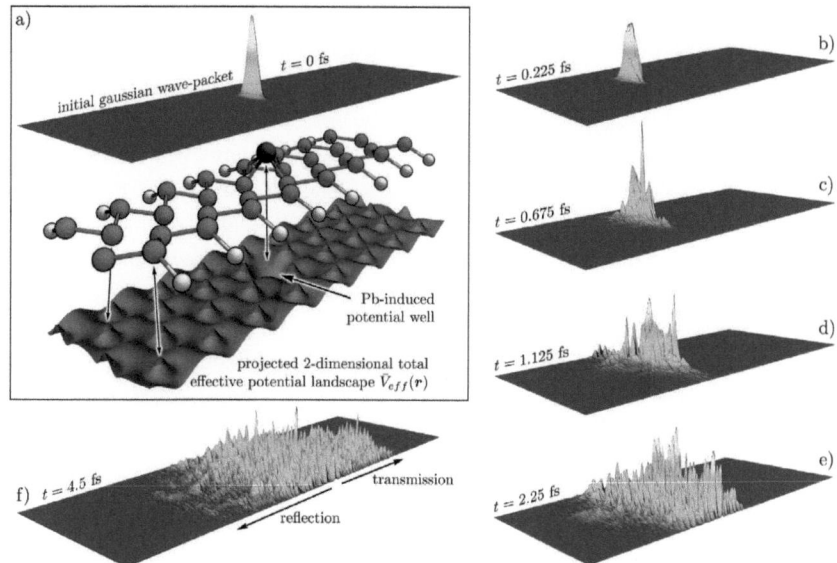

Figure 3.9: *a) Computational setup for numerically solving the 2-dim Schrödinger equation for a gaussian wave-packet of Fermi-energy electrons traveling towards the Pb-induced potential well. b-f) Snapshots of amplitude distribution after 0.225, 0.675, 1.125. 2.25 and 4.5 fs, respectively. The scaling is adapted with time for optimum viewability.*

As illustrated in Fig. 3.10, the systems studied here differ drastically with respect to the local potential. A very deep potential well is formed upon Pb adsorption, while additional In atoms barely change the potential along the wire direction. Thus it seems likely that the conductance modification observed upon Pb adsorption obeys a similar mechanism as proposed in the case of CO adsorbed Au chains [190]. Judging from Fig. 3.10, however, this mechanism cannot explain the conduc-

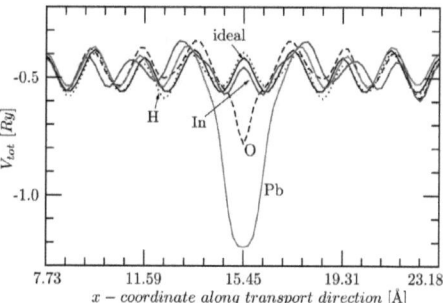

Figure 3.10: *Averaged (see text) effective potential along the wire direction calculated for ideal and adatom-modified In/Si(111) structures (Publ. [11]).*

tance drop upon In deposition: In adatoms reduce the In nanowire transmittance even more than Pb, but do not give rise to large potential fluctuations.

These considerations are corroborated by 1-dimensional model calculations, where the time-dependent Schrödinger equation was solved numerically for the potentials shown in Fig. 3.10 (cf. section 2.2.1a). The transmission of Fermi wave-vector electrons across the Pb potential well is reduced by 8%, while a reduction by only 3 % is obtained in case of In. The difference becomes even more pronounced if the Schrödinger equation is solved in 2 dimensions. Fig. 3.9a shows a 2-dimensional projection of $V_{eff}(r)$ onto the plane of the Si(111) surface. Analogous to Fig. 3.10 the potential was extracted from a volume slice containing the In nanowire atoms as well as the adatoms within their covalent radii. A rotationally symmetric gaussian wave-packet was placed in front of the defect. Fig. 3.9b-f shows the time-development of the wave-packet. After an elapsed simulation time of 4.5 fs integrating over the transmitted and received amplitude distributions yields a 12% conductance drop for Pb as opposed to 2% for In.

4.3.3 Structural effects

In order to understand the even more pronounced conductance drop upon In deposition, the initial observation should be remembered that the adatoms deform the nanowire to different degrees, as can be seen from Tab. 1.1 and Figs. 1.3 and 1.4. While the smallest deformations are observed for H and Pb adsorption, In causes substantial strain. Oxygen represents an intermediate case. The computational modeling allows for separating the impact of the adatom-induced structure deformation from the impact of the adatom itself: Fig. 3.11 shows the quantum conductances resulting from transport calculations that were performed for nanowire structures that are deformed according to their relaxation in response to the adatom, but, however, do not contain the adatom itself. The results are compiled in Tab. 3.3. As can be seen here, the – comparatively small – geometry changes of the In nanowire upon adsorption of H or Pb do not substantially reduce the wire conductance. This is in contrast to the stronger distortions caused by the adsorption of oxygen or indium, where moderate

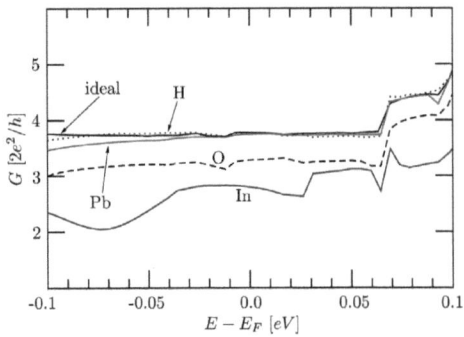

Figure 3.11: *Quantum conductance spectra for electron transport along the wire direction calculated for ideal and adatom-modified In/Si(111) structures that do, however, not contain the adatom itself (cf. Publ. [11]).*

4.3. Conductance quenching mechanisms

Adatom species	$G\ [\frac{2e^2}{h}]$	$G'\ [\frac{2e^2}{h}]$
Ideal	3.75	3.75
H	3.60	3.74
O	2.43	3.25
In	2.31	2.77
Pb	3.11	3.72

Table 3.3: *Average quantum conductance G in the energy interval ± 0.05 eV around the Fermi energy. G' refers to the respective adatom structure without the adatom.*

to strong conductance reductions are calculated, respectively. These results are suitable to explain the conductance drop observed in case of In deposition and, in part, for O deposition as well.

4.3.4 Discussion

The results presented above now allow for a classification of the adatom-induced conductance modifications:

- Pb adsorption does not substantially modify the nanowire geometry, but forms a deep potential well that effectively scatters the electrons and thus reduces the transmittance.

- No significant potential well forms upon In deposition. Here the conductance drop is related to the wire deformation.

- Both factors contribute in the case of O. A moderate potential well is formed and on top of that the In nanowire gets somewhat deformed. Both effects combined result in a conductance drop of similar magnitude as calculated upon In deposition.

- Hydrogen, finally, does neither act as a potential well nor does it significantly strain the nanowire geometry. Consequently, it has no substantial impact on the electron transport. This is confirmed by the conductance measurements in Ref. [58].

The approach to explain the conductance quenching by modifications of the local density of states did not yield quantifiable results. However, as a qualitative argument from an illustrative point of view the local DOS approach is not without merit. Fig. 3.12 shows the local DOS of clean and Pb-adsorbed In nanowires. While the ideal In

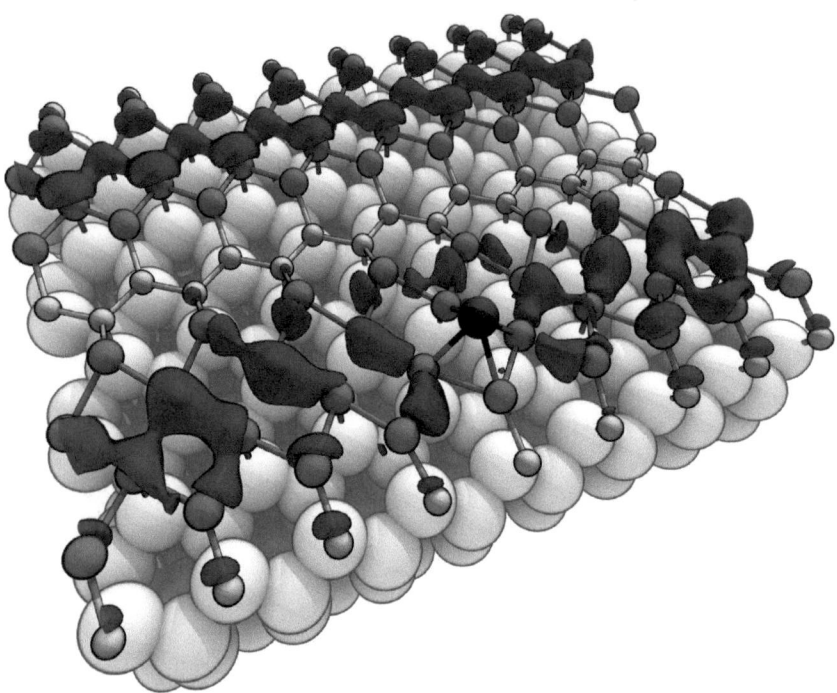

Figure 3.12: *Isodensity surface of the local DOS at E_F, illustrating the local DOS at the ideal In nanowires and its modification upon Pb adsorption.*

nanowire exhibits a single continous channel between the In zigzag chains, a strong modification and depletion around the Pb adsorption site is observed. Thus the formerly continous conduction channel is effectively severed. The results presented in this chapter were published in Publ. [11].

Nature uses only the longest threads to weave her patterns, so that each small piece of her fabric reveals the organization of the entire tapestry.

– Richard Feynman

CHAPTER 5
Optical properties

Ever since the discovery of the (4×1)→(8×2) phase transition by Yeom *et al.* more than 10 years ago [30] the structure and properties of the low temperature (LT) phase have remained controversial (cf. chapter 1.2.1). As discussed in chapter 3 and Publ. [14] an unambigous identification of the internal structure of the LT phase on the basis of the total energy is problematic at best. The energy differences between the competing structures are very small and depend on numerical details as well as the approximations made in the calculations concerning, e. g., the treatment of the exchange and correlation effects and the In $4d$ states.

Given the ambiguities of the total-energy calculations in determining the ground-state structure of the (8×2) LT phase, the comparison of optical fingerprints calculated for structural candidates with measured data may be helpful. In this chapter it is demonstrated that the comparison of calculated and measured *reflectance anisotropy spectroscopy* (RAS) data in the mid-infrared regime is suitable to settle the more than 10 year old discussion about the nanowire's true ground state.

5.1 Optical anisotropy in the visible spectral range

The calculated electron transport properties considered in chapter 3 give strong evidence for hexagon formation. On the other hand, the previously calculated optical anisotropies [52] for the (4×1) and the trimer-reconstructed (8×2) nanowire arrays are compatible with the measured evolution of the surface optical response during the (4×1)→(8×2) phase transition [51]: reflectance anisotropy spectroscopy (RAS) at the Si(111)-(4×1)-In surface [49, 50, 191] shows an optical anisotropy in the energy region of 2 eV, which splits into two peaks, at 1.9 and 2.2 eV, upon formation of the LT phase of the Si(111)-In surface [51, 192]. In order to help clarifying the structure of the In nanowire array, the optical anisotropies were calculated for the (4×1) surface and the two structural models suggested for the LT (8×2) phase, i. e. the trimer and the hexagon model.

5.1.1 Computational details

The following results are obtained within the local density approximation (LDA) for exchange and correlation as implemented in VASP [104], with the electron-ion interaction described by projector-augmented wave (PAW) potentials. The In 4d states are treated as core electrons. Unlike in the previous chapters, the In/Si(111)-(4×1) and (8×2) surfaces are simulated by *symmetric* slabs with 12 Si bilayers (cf. 2.1.5b) and a vacuum region equivalent in length. The k-space integrations are performed employing uniform meshes equivalent to 64 and 960 points in the (1×1) surface Brillouin zone for electronic structure and optical response, respectively. As discussed in chapter 2.3.1 the reflection anisotropy $\Delta R/R$ for light polarized along i and j can be derived from a slab calculation and is given by

$$\frac{\Delta R}{R}(\omega) = \frac{2\omega d}{c} \Im \left\{ \frac{\epsilon_{ii}^{slab}(\omega) - \epsilon_{jj}^{slab}(\omega)}{\epsilon_b(\omega) - 1} \right\} \quad (1.1)$$

where $\epsilon_b(\omega)$ is the bulk dielectric function and $\epsilon_{ij}(\omega)$ is the dielectric tensor of the slab with thickness d. In the following the polarization directions are assumed to be the [11$\bar{2}$] and [110] directions, respectively.

The dielectric tensor is obtained within the independent-particle approximation based on the electronic structure calculated within DFT-LDA. A scissors operator is used to correct the band-gap underestimation. Thereby a constant shift of 0.5 eV is applied to all unoccupied DFT eigenvalues that lie more than 0.5 eV above the highest occupied electronic state. This ensures an approximate reproduction of the measured Si bulk band structure. For the energetically lower lying unoccupied states a linear interpolation up to a zero shift for states directly at the Fermi energy is used in order to model the energy-dependence of the electronic self-energy. Beyond this approximation of the self-energy, many-body effects such as excitonic and local-field effects are completely neglected. While this certainly results in a loss of quantitative accuracy and may strongly affect the calculated line shape, past experiences indicate that the calculated spectra can still be expected to be qualitatively reliable. As a consequence of the substraction $\epsilon_{ii}^{slab}(\omega) - \epsilon_{jj}^{slab}$ in Eq. (1.1) RAS calculations profit from a strong degree of systematic error cancellation. This holds in particular if differences and trends for very similar systems are considered, as in the present case.

5.1.2 Results

The calculated optical anisotropy is shown in Fig. 1.1. For all structural models the RAS is negative for nearly the complete energy range considered. The spectra are dominated by a very strong anisotropy around 2 eV photon energies of about 3.6% and 4.8% for the (8×2) and (4×1) surface reconstructions, respectively. This is in excellent agreement with the experimental findings [49, 50, 191] of a very pronounced anisotropy of 2.4–4.0% (if the intensities are considered) at 2 eV. The fact that the calculated optical anisotropy is somewhat larger than measured is expected. These results refer to a single-domain Si(111) surface free of steps and other defects that is covered with a perfect In nanowire array. Defects in the In chains and signals from minority domains present at the real surface may reduce the peak at 2 eV.

The optical anisotropy at 2 eV is necessarily related to surface states, because this energy is far below the direct optical gap of bulk Si. However, it is not directly related to the metallicity of the nanowires, which at low frequencies is expected to result in a stronger optical coupling for light polarized along the chain direction rather than perpendicular to the chains. Therefore, the quasi-one-dimensional metallicity of the In chains should lead to positive optical anisotropies, which are indeed

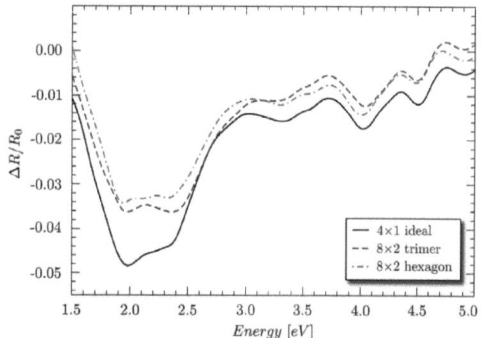

Figure 1.1: *RAS spectra calculated for the (4×1) ideal and (8×2) trimer and hexagon models of the In/Si(111) nanowire array, respectively (cf. Publ. [9]).*

observed for photon energies below 1 eV [51]. This energy region is addressed specifically in the following section 5.2.

Performing the identical RAS calculations at the Γ-point only yields qualitatively the same spectra. This allows for an easy identification of the transitions exhibiting a strongly anisotropic behaviour. The negative anisotropy around 2 eV is related to a multitude of optical transitions perpendicular to the In chain direction that are either related to pure In chain states or involve the In-Si bonds. The former occur around 1.9 eV whereas the latter are observed at energies slightly above 2 eV. This shoulder was not seen in previous calculations. It turns out that the optical response is extremely

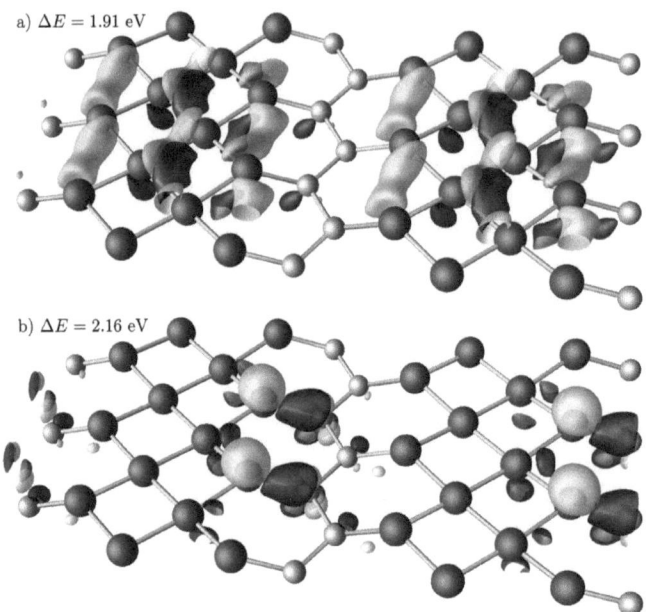

Figure 1.2: *Squared wavefunctions of surface electronic states that contribute strongly to the optical anisotropy of the In/Si(111)-(4×1) nanowire array. The upper/lower panel shows the valence (blue) and conduction (white) states that cause an anisotropy at 1.9/2.2 eV. The isosurfaces are drawn for a density of 0.003 Å^{-3} (cf. Publ. [9]).*

sensitive with respect to structure: The (4×1) surface must be relaxed up to remaining forces of lesser than 0.02 eV/Å for the shoulder to develop. The shoulder is also absent in the measurements, likely due to thermal broadening. Fig. 1.2 shows two representative transitions for the case of the (4×1) reconstructed nanowire array.

The RAS of the hexagon model for the (8×2) LT phase has not been calculated before. In order to assist in the identification of structural models the RAS was calculated for both the hexagon and the trimer model, see Fig. 1.1. The spectra are similar to the one calculated for the (4×1) phase. However, the shoulder obtained for the (4×1) phase at around 2.4 eV develops into a separate peak for both models. Also, the optical anisotropy is reduced from about 4.8–3.6%. The splitting of the 2 eV anisotropy into two separate peaks corresponds exactly to the optical signature of the (4×1)→(8×2) phase transition found experimentally [51, 192]. In addition to the peak splitting, the calculated minimum of the RAS shows a slight redshift by about 0.1 eV, which

5.1. Optical anisotropy in the visible spectral range

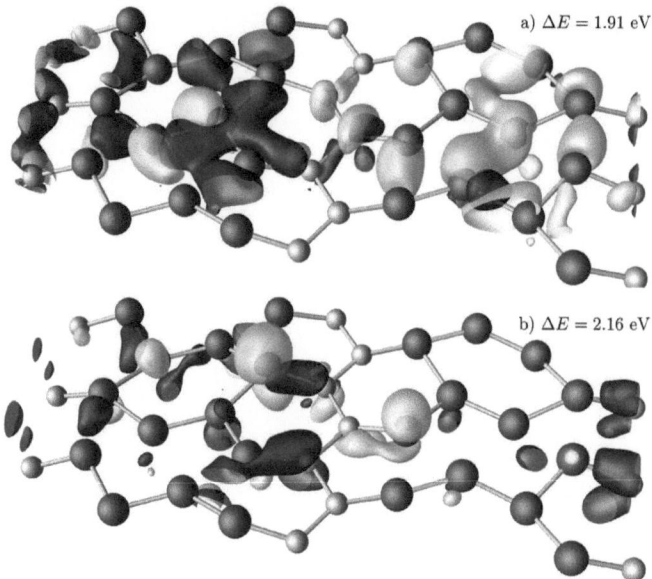

a) $\Delta E = 1.91$ eV

b) $\Delta E = 2.16$ eV

Figure 1.3: *Squared wavefunctions of surface electronic states that contribute strongly to the optical anisotropy of the hexagon model for the In/Si(111)-(8×2) nanowire array. The upper/lower panel shows the valence (blue) and conduction (white) states that cause an anisotropy at 1.9/2.2 eV. The isosurfaces are drawn for a density of 0.002 Å^{-3} (cf. Publ. [9]).*

again is in very good agreement with the experimental findings that state a shift from 1.96 to 1.90 eV during the phase transition. Fleischer *et al.* [51] argue that the measured changes in the optical anisotropy cannot be explained as a temperature-induced sharpening of the original 2 eV peak, since the overall width of the measured structure is much larger for the LT phase. The changes of the RAS spectra can thus only be explained by electronic and structural modifications of the In nanowire array accompanying the phase transition. Surprisingly, both the hexamer and the trimer model for the (8×2) phase give rise to a very similar optical anisotropy. Within the numerical accuracy of the present calculations a discrimination of the models with respect to the degree of agreement with experiment is not possible in the *visible* spectral range.

Due to the fact that a multitude of electronic surface states is relevant for the occurence of the optical anisotropy characteristic for the nanowire array – in fact about 20 transitions are found to be relevant – it is impossible to provide a simple picture for the spectral changes accompanying the phase transition. The symmetry reduction upon hexagon or trimer formation changes the orbital character of nearly all In chain related states and leads to a reduction of the oscillator strength of many transitions responsible for the overall larger optical polarizability perpendicular to the In chain direction. As an example, Fig. 1.3 shows the electronic states responsible for two optical transitions that lead to strong anisotropies for the case of the (8×2) hexamer-reconstructed nanowire array. The orbital character of these states is somewhat similar to the case of the ideal (4×1) nanowire array (cf. Fig. 1.2) and the respective optical transitions occur at similar energies. The transition matrix elements, however, are smaller by about one third (cf. Publ. [9]).

5.2 Structure determination by mid-infrared response

As discussed in the previous section, conventional RA techniques in the visible range of the spectrum do not allow to distinguish between the competing structural models for the (8×2) LT phase. However, these results represent a very solid foundation allowing to proceed to the calculation of the RAS in the mid-infrared (IR) regime. The small differences in geometry of the trimer and hexagon model lead to significant changes in the band structure near the Fermi level (cf. chapter 3.2). Direct optical transitions in this energy region, as probed by infrared reflection anisotropy spectroscopy (IRAS), are expected to be very sensitive to such changes. Both IRAS *ab initio* calculations and measurements are presented in Publ. [3,8] for the first time for this system. These results are suitable to resolve the 10 year old discussion about the In/Si(111) nanowires' true ground state.

5.2.1 Computational details

Since the (4×1) ideal nanowire and the (8×2) trimer model are metallic in nature, they feature *intraband* transitions in addition to the usual interband transitions (cf. chapter 2.3.2b). Below transition energies of about 1 eV the spectra are increasingly dominated by intraband transitions. Thus to calculate the IRAS intraband transitions need to be taken into account as well.

The calculation of the intraband contributions to the dielectric tensor is performed by introducing a small q-vector, with the optical transitions proceeding along $k \to k + q$. The Drude contribution from the intraband transitions is then obtained in the limit $q \to 0$. In the numerical evaluation of the intraband contributions the k-point sampling and the chosen q-vector become two extremely critical convergence parameters.

5.2. Structure determination by mid-infrared response

A sufficiently small q-vector must be ensured to reproduce the $q \to 0$ limit. However, this simultaneously enforces a correspondingly dense k-point mesh. In the present work the k-space integrations to calculate the intraband contributions is performed using a uniform mesh equivalent to 30,720 points in the (1×1) surface Brillouin zone. The chosen q-vectors are $\|q\| = 0.0125$ and $\|q\| = 0.0417$ along and perpendicular to the nanowire direction, respectively. Due to the enormous computational requirements these calculations could only be performed for the ideal In/Si(111)-(4×1) reconstruction and not for any larger reconstructions. However, at a later stage it will become clear that intraband contributions can be safely neglected for the (8×2) trimer model.

For the numerical calculation of the intraband contributions the DP code [182] is employed. Subsequently the intraband contributions are separated and added to the dielectric tensors obtained by VASP, where the intraband contributions had been neglected. This ensures an optimum comparability between the spectra. In the case of the semiconducting (8×2) hexagon model the numerical details are identical to the ones in the previous section, except the quasiparticle corrections: states close to the Fermi energy – as probed by IRAS – are shifted very little by quasiparticle corrections. A simple linear scissors operator typically does not improve the accuracy of the resulting spectra. Unfortunately, the (8×2) cell is prohibitively large for many-body perturbation calculations of the required accuracy. Therefore no quasiparticle shifts are employed for any of the following spectra.

5.2.2 Mid-infrared optical anisotropy

Fig. 2.4 shows the measured and calculated IRAS spectra. The experimental data reveal a dramatic difference between the RT (4×1) and LT (8×2) phases. There is a smooth Drude-like increase to lower energies without any interband transitions remaining below 1 eV for the RT (4×1) phase. This is in contrast to the behaviour above 1 eV, where the optical response is dominated by interband transitions. The larger Drude-like response parallel to the chains arises from the highly anisotropic conductivity of this surface, as measured by four-point probe STM [43, 44].

The LT (8×2) phase shows two sharp positive peaks at 0.50 and 0.72 eV, while the low energy Drude tail is removed. Its replacement by these peaks indicates a metal-insulator (MI) transition. There is no evidence of residual metallic behaviour in the (8×2) phase in the mid-IR regime. Positive anisotropy indicates that optical transi-

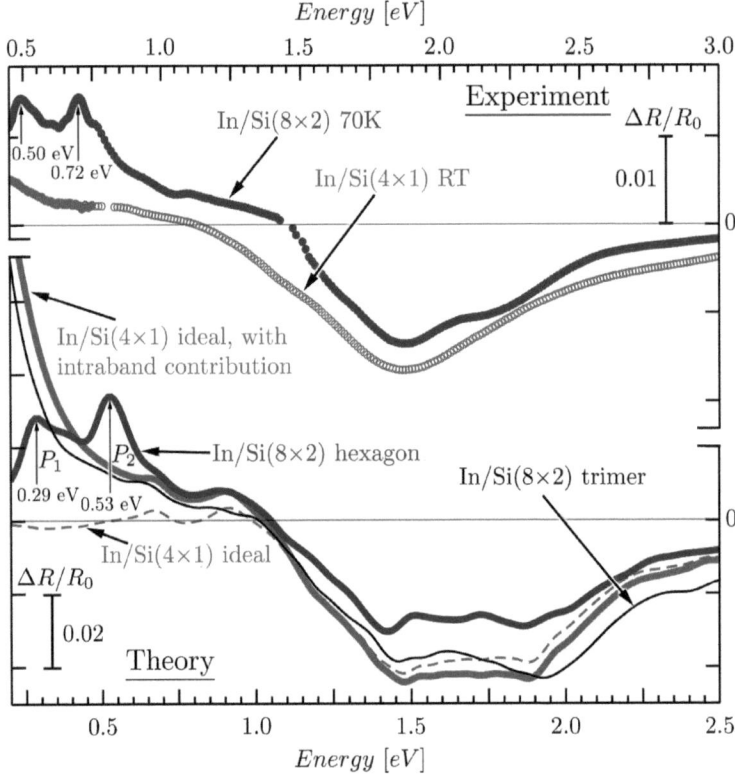

Figure 2.4: *RA-spectra of Si(111)-(4×1)In at RT (300K) and Si(111)-(8×2)In at LT (70K): upper, experiment; lower, theory. Note the different scales (cf. Publ. [3,8]).*

tions parallel to the chains are dominant in this spectral region. Both phases show the broader negative 1.9 eV feature, which splits below the MI transition. The calculated anisotropy of the ideal (4×1) reconstruction agrees well with experiment, provided intraband transitions are taken into account. It should be noted that the calculated Drude-like features turn out to be apparently too pronounced. This is due to the different scales in Fig. 2.4. The Drude tail is inherently unaffected by any quasiparticle shifts, while the interband transitions are redshifted due to the absence of quasiparticle corrections. This results in a "compression" of the spectra towards lower energies. Comparing the Drude-like features on an absolute scale, i. e. at 0.5 eV, yields a good agreement with experiment.

5.2. Structure determination by mid-infrared response

While no intraband transitions were calculated for the trimer model of the (8×2) LT phase, the RAS still shows a steep rise in analogy to the Drude-like behaviour of the (4×1) RT phase. This results from an increased density of states near the Fermi level due to band folding effects (cf. chapter 3.2). However, since this model exhibits only one band crossing the Fermi level – with rather low dispersion in comparison to the (4×1) RT phase – the contribution of intraband transitions can be expected to be almost negligible for the trimer model.

The calculated anisotropy of both the trimer and hexagons models of the (8×2) structure agree well with experiment above 0.7 eV, as discussed in the previous section. Below 0.7 eV, only the hexagon model looks similar to the experimental results. In particular, two positive peaks are predicted, separated by 0.24 eV. This splitting agrees very well with the experimental splitting of 0.22 eV. Detailed comparison reveals that the calculated mid-IR peaks are redshifted by about 0.25 eV (note the different scales in Fig. 2.4). The underestimation of excitation energies is typical for DFT calculations where self-energy effects are neglected. As mentioned previously, the complexity and size of the In nanowire structure prevents the calculation of optical spectra employing many-body perturbation theory that includes self-energy and excitonic effects [193]. Quasiparticle calculations for the high-symmetry points of the hexagon model surface band structure found self-energy effects to increase the lowest transition energies by about 0.26 eV on average (cf. Publ. [14]). A larger shift of 0.5 eV, typical for Si excitation energies [193], applies to the higher energy negative optical anisotropies, because the optical transitions involve Si states (cf. Publ. [9]). Allowing for these energy shifts, the agreement between the calculated and measured RAS spectra is truly impressive.

5.2.3 Transitions responsible for the observed anisotropy

Identifying the origin of the two peaks in the mid-IR regime is not as easy as tracing the transitions responsible for the strong negative anisotropy around 2 eV. In contrast to the 2 eV feature there is no single k-point in existence that even qualitatively yields the two peaks. However, according to Eq. (3.220) in chapter 2.3 the dielectric tensor $\epsilon_{ij}(\omega)$ is given by a summation over the k-points and the bands. The additive nature of $\epsilon_{ij}(\omega)$ allows to implement a suitable search algorithm in a straight forward way. Searching through all direct transitions at every k-point yields mainly optical transitions close to the X and M points of the surface Brillouin zone, indicated by P_1 and P_2/P_2' in Fig. 2.5, corresponding to the labels in Fig. 2.4. Around $\overline{XM}$ nearly parallel valence and conduction bands close to the Fermi level give rise to a high joint density

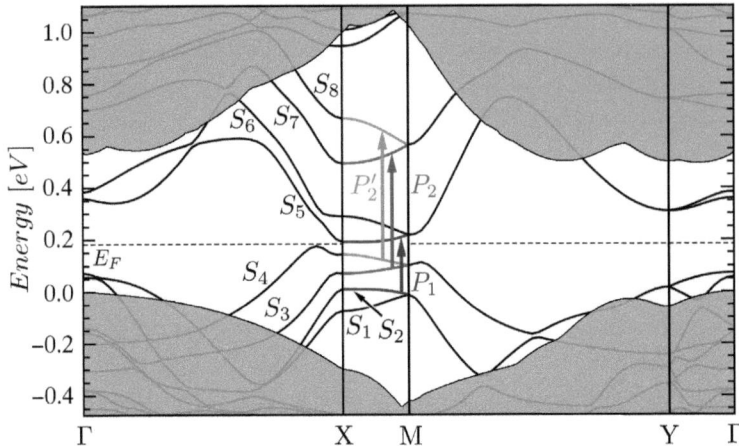

Figure 2.5: *Band structure of the hexagon model for In/Si(111)-(8×2) calculated within DFT-LDA. Pronounced optical transitions showing up in the RAS spectra as the peaks P_1 and P_2 in Fig. 2.4 are marked. Gray regions correspond to the projected Si bulk bands. The bulk valence band maximum is chosen as energy zero. The Fermi level is indicated (cf. Publ. [3,8]).*

of states. The corresponding surface electronic states at the X and M high-symmetry points of the surface Brillouin zone are shown in Fig. 2.6.

However, not all of the relevant transitions arise from the region of the $\overline{XM}$ high-symmetry line, but from inside the Brillouin zone as well. Thus the common practice of drawing the bands only along the high-symmetry lines is insufficient in this case. To visualize the relevant transitions a 3-dimensional (3D) representation of the entire surface Brillouin zone is shown in Fig. 2.7. The notation of the surface bands $S_1 - S_8$ and the peaks P_1, P_2/P_2' refers to Fig. 2.5 and 2.4, respectively. The transitions indicated in Fig. 2.7 b/c) take place over the whole width of the surface Brillouin zone, while the transitions in d/e) occur either near the $\overline{\Gamma X}$ or $\overline{MY}$ high-symmetry lines only. It can also be seen that in some cases transitions from and to the same bands contribute to different peaks, as shown in Fig. 2.7e). Transitions from $S_2 \rightarrow S_5$ contribute to either P_1 or P_2, depending on the exact location of the transition inside the surface Brillouin zone.

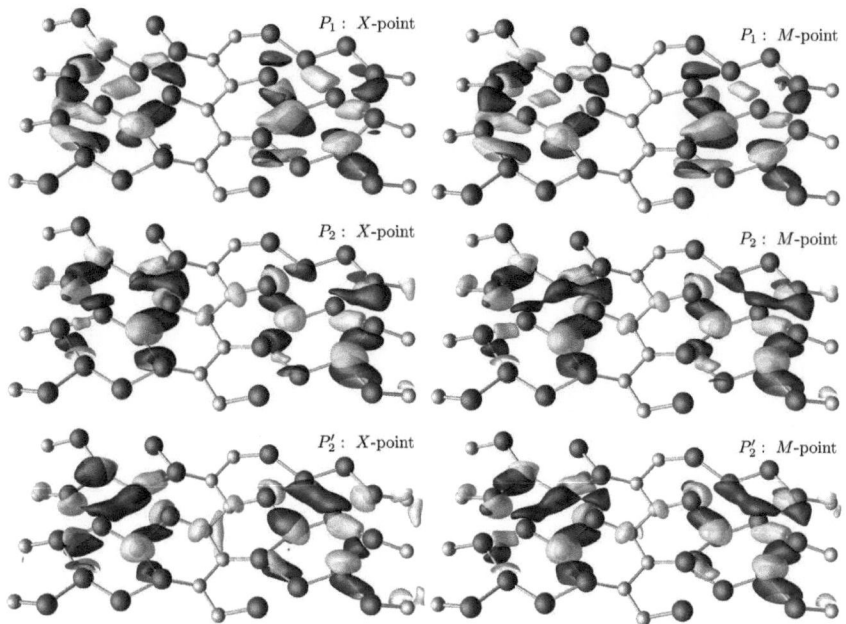

Figure 2.6: *Squared wavefunctions of surface electronic states at the X and M high-symmetry points that contribute strongly to the optical anisotropy of the hexagon model for the In/Si(111)-(8×2) nanowire array in the mid-IR regime. The notation is consistent with Figs. 2.4-2.7. The isosurfaces are drawn for a density of 0.0015 Å^{-3}. Blue and white isosurfaces correspond to valence and conductance states, respectively.*

5.3 Discussion

In conclusion, the optical anisotropies for ideal and hexagon as well as trimer reconstructed models for the In/Si(111) nanowire array have been calculated from *first principles*. The comparison with measured data shows that above 0.7 eV – as measured by conventional RAS – the optical signatures of both the hexamer as well as the trimer model are very well suitable to explain the spectral changes acquired during the formation of the LT phase of the In/Si(111) surface. Predictions of the reconstruction model are impossible in this spectral range. Since the most pronounced changes between the band structures of competing structural models occur near the Fermi energy, the mid-IR regime is expected to be more revealing.

Extending the optical calculations towards lower energies the first mid-IR *ab initio*

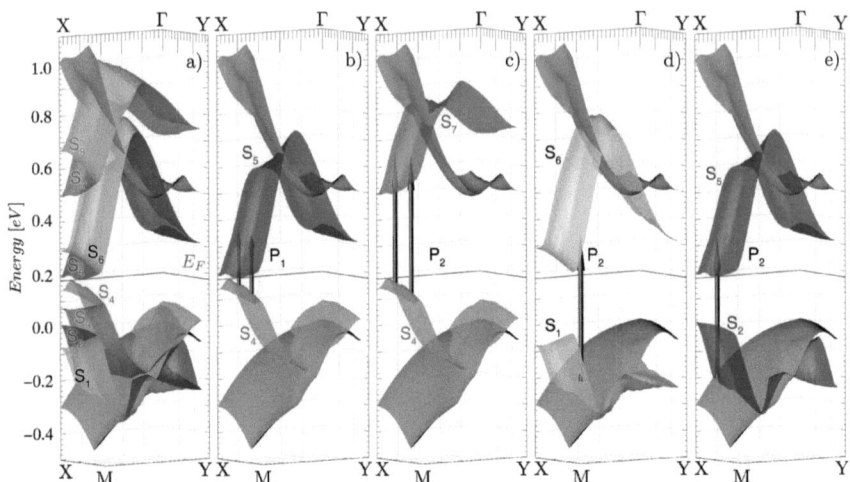

Figure 2.7: *a) 3-dimensional band structure over the entire surface Brillouin zone of the hexagon model for In/Si(111)-(8×2) calculated within DFT-LDA. The notation of the bands is consistent with Fig. 2.5. Pronounced optical transitions from within the Brillouin zone causing the peaks P_1 and P_2 in the RAS spectra are shown in b) and c-e), respectively. The Fermi level is indicated by E_F. (cf. Publ. [3]).*

optical response calculations including intraband transitions have been reported for the In/Si(111) nanowire system. In comparison with recent experimental data, these optical response calculations reproduce the measured features for the hexagon model. While excluding the trimer model these results provide strong evidence in favor of the hexagon model. They are thus suitable to settle the 10 year old dispute about the In nanowire ground state. The work presented above resulted in a joint experiment-theory publication (Publ. [8]).

A little wisdom is no doubt possible; but I have found this happy certainty in all things: that they prefer – to dance on the feet of chance.

– Friedrich Nietzsche

Chapter 6
Thermal properties

In the previous chapter it was demonstrated, that only the hexagon model for the (8×2) LT phase can explain the observed spectroscopic properties. While these data effectively confirm the hexagon ground state model, the question regarding the phase transition's driving mechanism is still open. Both the In/Si(111)-(4×1)/(8×2) surface's thermal properties as well as the phase transition itself have been studied previously by *Raman spectroscopy* (RS) [46] and temperature-dependent transport measurements [33, 44]. However, a detailed theoretical understanding of these data has yet to be achieved. The present chapter is dedicated to the calculation of the phonon modes and their comparison with the Raman spectroscopy data from Ref. [46]. Based on these data several soft phonon modes could be identified as the phase transition's driving mechanism. In conjunction with large-scale MD calculations an approach for the calculation of the temperature-dependent transport properties is presented.

6.1 Phonon spectra in theory & experiment

6.1.1 In/Si(111)-(4×1) surface

Beginning with the In/Si(111)-(4×1) surface the phonon modes are calculated within DFT-LDA as implemented within the VASP package. The In $4d$ states are frozen into the core. A plane-wave cutoff of 250 eV and a uniform mesh equivalent to 128 k-points in the (1×1) surface Brillouin zone were found sufficient to lead to well-converged results. The (4×1) surface is simulated by repeated asymmetric slabs with 3 Si bilayers and a vacuum region equivalent in length. Since Raman spectroscopy measures only modes near the Γ-point of the surface Brillouin zone the calculation was restricted to zone center phonon modes in the (4×1) unit cell. The modes were obtained within the frozen-phonon (FP) approach as described in section 4.2.3, choosing a displacement of 0.1 Å. This is somewhat larger than usual. However, test calculations employing a

Character	A'	A''
In surface mode	51, 53, 67, 85	17, 28, 66
In/Si surface mode	71, 100, 121, 141, 145	70
In/Si resonance mode	76, 198, 204, 265, 394	64
Si/Si resonance mode	279, 421 448, 462, 475	114, 119, 127
Si surface mode	259, 269, 317, 323, 375	95, 436

Table 1.1: *Calculated frequencies of A' and A'' surface vibrational Γ-point modes of the In/Si(111)-(4×1) surface in cm^{-1}. Modes that are localized mostly within the top atomic layer are termed surface modes, while the others are denoted as resonance modes in accordance with Ref. [47]. Any modes that are not localized within the upper two atomic layers are discarded. The green/violet color scheme refers to the modes marked in Fig. 1.1.*

displacement of 0.05 Å yielded results that were identical within the numerical accuracy. In section 6.2.5 it turns out that for the temperature range of interest the average elongation is of the order of 0.1 Å. Hence this displacement is adopted for all following frozen-phonon calculations.

The calculated Γ-point phonons can be classified according to the irreducible representations of the point group of the system. As the In/Si(111)-(4×1) surface features a $1m(C_s)$ symmetry, the only symmetry operation applicable – besides the translations of the 2-dimensional lattice – is related to the $[1\bar{1}0]$ mirror plane. The normal direction of this mirror plane is located parallel to the In chain direction. Modes whose elongation patterns are situated within and perpendicular to the $[1\bar{1}0]$ mirror plane are labeled A' and A'', respectively.

The obtained modes are categorized and compiled in Tab. 1.1. Their character has been determined according to their displacement pattern, not according to the occurence of the modes' frequencies within pockets of the projected bulk phonon branches. As Raman spectroscopy probes only surface phonons, any phonons that are not localized within the uppermost two atomic layers are excluded. Thus 34 surface modes out of 42 total remain. For each mode Fig. 1.1a/c shows the degrees of localization at the In atoms and the top atomic layer, respectively. In between in Fig. 1.1b a comparison of the calculated and measured modes (measurements taken from Ref. [46]) is presented. Not all of the calculated modes are observed experimentally, probably due to selection rules or low scattering efficiency in the experiment. It was tried to assign the theoretically derived modes to the experimentally observed ones, as indicated by vertical lines. The single strong A'' mode at 28 cm^{-1} is recognized easily in the calculated spectra. Besides this mode 7 other A'' modes are present in

6.1. Phonon spectra in theory & experiment

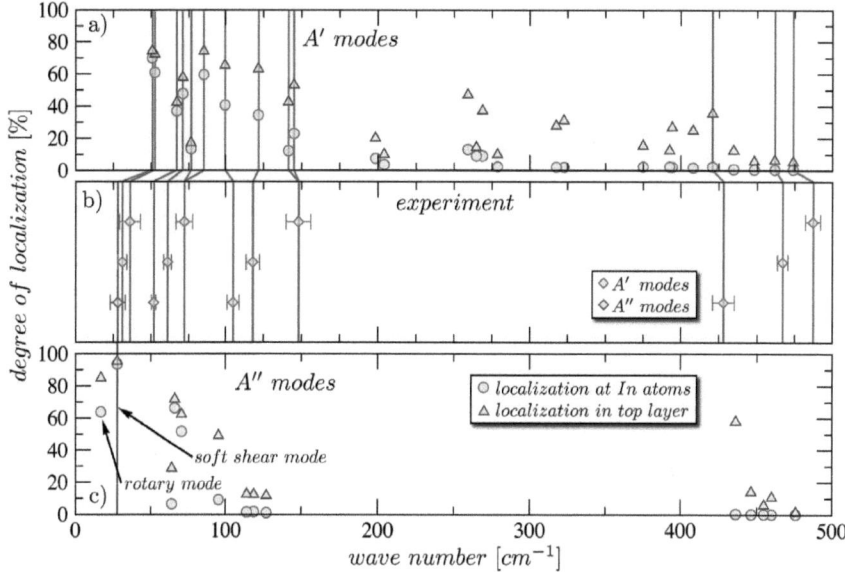

Figure 1.1: *a/c) Frequencies and degrees of localization for the calculated Γ-point phonon modes of the In/Si(111)-(4×1) surface. b) Comparison with the measurements in Ref. [46]. Error bars indicate the measured line widths. Experimentally observed modes are marked by* green *and* violet *lines for A′ and A″ modes, respectively. The commented modes refer to Figs. 1.2 and 1.3.*

the theoretical spectra. However, none of them is observed experimentally. Up to a wave number of 150 cm^{-1} 9 A′ modes are obtained, exactly as many as measured. All of them are localized to a high degree in the top atomic layer. Only the modes in the range of 100-150 cm^{-1} correspond closely to the measured wave numbers. The 5 modes between 50 and 85 cm^{-1} are blueshifted in comparison by about 13 cm^{-1}. Fleischer et al. noted that the modes observed at wave numbers of 72 and 148 cm^{-1}, respectively, are possibly two modes very close in energy. The calculated modes confirm this to be the case for the 148 cm^{-1} mode, but not the 72 cm^{-1} mode.

Fleischer et al. [46] suggested that soft-shear distortions as proposed by González et al. [35, 37] might provide an explanation for the single broad mode of A″ symmetry observed at 28 cm^{-1}. A closer look at the eigenvectors of the A″ mode at 27 cm^{-1} obtained from the present calculations is quite revealing: this mode features indeed a strong shear-distortion that displaces the In chains relative to each other forward and backward along the chain direction (cf. Fig. 1.2a). With a wave number of 28 cm^{-1} = 3.43 meV this mode is also sufficiently soft to be occupied near and below the

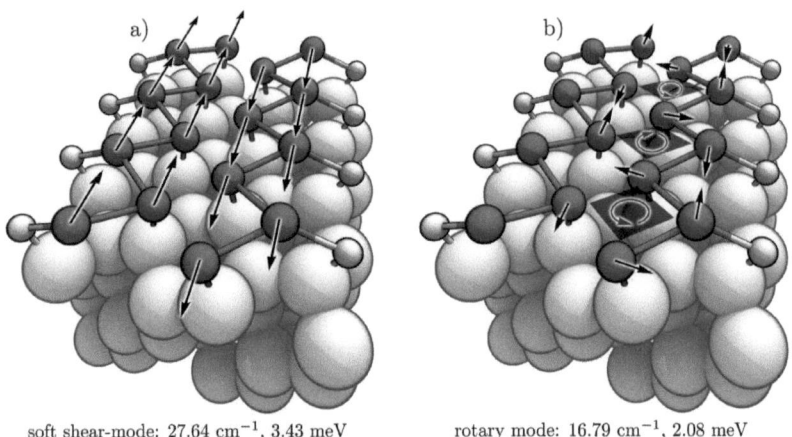

soft shear-mode: 27.64 cm^{-1}, 3.43 meV   rotary mode: 16.79 cm^{-1}, 2.08 meV

Figure 1.2: *Eigenvectors of the **a**) soft shear and **b**) rotary modes, respectively. The circular blue arrows indicate the rotation direction of the rectangle constituted by the 4 adjacent In atoms. These modes are candidates for inducing the shear and trimerization displacements that are required for hexagon formation in the LT phase. They are sufficiently low in energy to remain excited below the critical temperature $T_C = 120$ K $= 10.34$ meV.*

phase transition's critical temperature $k_B \cdot T_C = k_B \cdot 120$ K $= 10.34$ meV. It can thus be considered a candidate for triggering the phase transition by shearing the In chains, as required for hexagon formation.

However, besides the shear distortion both the hexagon and trimer models of the LT phase feature a trimerization of the outermost In chain atoms. Interestingly, the calculations predict another low energy mode at 17 cm^{-1} = 2.08 meV, where 4 adjacent In atoms form rectangles that rotate back and forth (cf. Fig. 1.2b). This mode could facilitate the trimerization during the formation of the LT phase. Naturally, while the eigenvectors illustrated in Fig. 1.2b support the trimerization of only one chain, there is a symmetrically equivalent degenerate mode for the other chain as well. However, these modes are X-point modes with respect to the (4×1) surface Brillouin zone and are thus not observed by Raman spectroscopy as performed in Ref. [46], at least not for structures with (4×1) translational symmetry.

This claim that the phase transition could be driven by a soft shear-mode and two degenerate soft rotary modes is supported by the fact that a linear combination of the frozen-phonon eigenvectors of the ideal In chains allows to reproduce both the hexagon and trimer models with high accuracy. To this end the frozen-phonon eigen-

6.1. Phonon spectra in theory & experiment

vectors of the (4×1) RT structure were calculated within the (4×2) unit cell. Subsequently, the resulting eigenvectors $u_{i,m}$ have been added to the positions x_i of the ideal structure according to:

$$x = \sum_{i,m} x_i^{ideal} + a_m \cdot u_{i,m} \qquad (1.1)$$

The indices i and m denote atomic and mode numbers, respectively. By optimizing the linear coefficients a_m both the trimer and hexagon models of the LT phase could be obtained within an accuracy of 0.01 Å average displacement per In atom.

The optimum linear coefficients for both models are shown in Fig. 1.3. The spectra are clearly dominated by the shear and rotary modes, respectively. Naturally, for the trimer model the shear contribution is negligible. While both models feature rather similar linear coefficients otherwise, the rotary contributions are also somewhat smaller for the trimer model. Besides the shear and rotary contributions, another mode is present that displaces the outermost In atoms along the normal direction, as illustrated by the inset in Fig. 1.3. The contribution of the remaining modes to the In atom displacement is rather small. If only the indicated shear, rotary and normal modes are taken into account both the hexagon and trimer structures are reproduced within an accuracy of 0.07 Å average displacement per In atom. With respect to the ideal chains the trimer and hexagon models feature an average displacement per In atom of 0.52 and 0.28 Å, respectively.

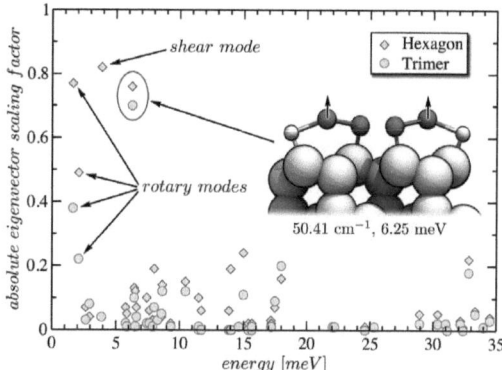

Figure 1.3: *Linear decomposition of the (4×2) hexagon and trimer models into phonon eigenvectors.*

Therefore both structural models for the LT phase can be largely reproduced by a linear combination of only the soft shear and rotary modes plus another mode inducing a normal displacement of the outermost In atoms. In combination with the at least local minima on the potential energy surface for the trimer and hexagon models these findings provide a strong indication that the phase transition is indeed phonon mode driven.

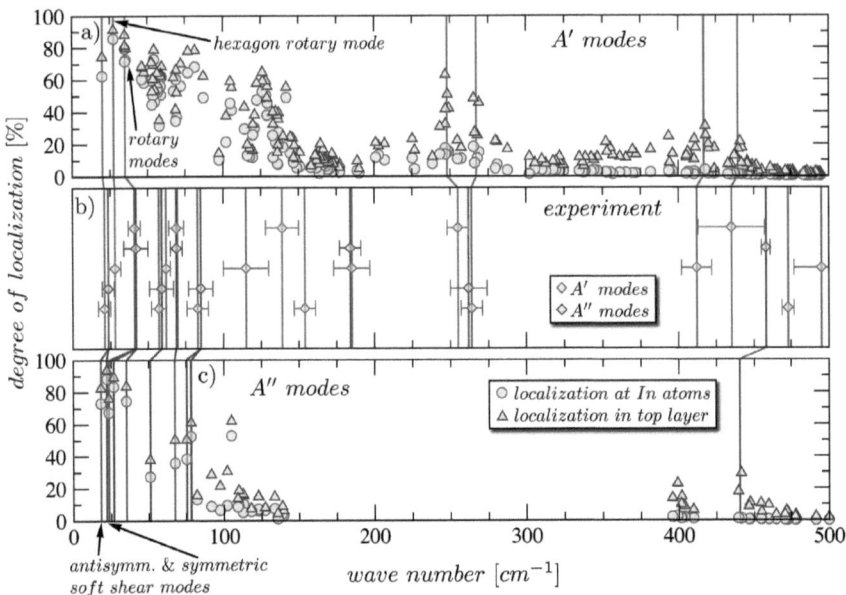

Figure 1.4: *a/c) Frequencies and degrees of localization for the calculated Γ-point phonon modes for the hexagon model of the In/Si(111)-(8×2) surface. b) Comparison with the measurements in Ref. [46]. Error bars indicate the measured line widths. Experimentally observed modes are marked by green and violet lines for A' and A'' modes, respectively. The commented modes refer to Figs. 1.5 and 1.6*

6.1.2 In/Si(111)-(8×2) hexagon structure

Turning to the (8×2) hexagon model for the LT phase, frozen-phonon calculations were performed in the (8×4) unit cell to obtain both Γ- and X-point phonon modes. While only the Γ-point modes are needed for comparison with the Raman spectroscopy results, the X-point modes are required for the later treatment of the transport properties in section 6.2. Due to the resulting size of the unit cell the amount of substrate had to be restricted to 2 Si bilayers. However, Raman spectroscopy measures only surface phonons and the transport properties depend very little on the substrate (cf. chapter 3.3.1). Test calculations in the (4×1) unit cell showed that the obtained frequencies and eigenvectors are barely affected as well. Otherwise the computational details are the same as in the previous section.

While the (4×1) RT phase features a $1m(C_s)$ symmetry, this is no longer the case for the (8×2) LT phase. Both trimer and hexagon formation break the mirror symme-

6.1. Phonon spectra in theory & experiment

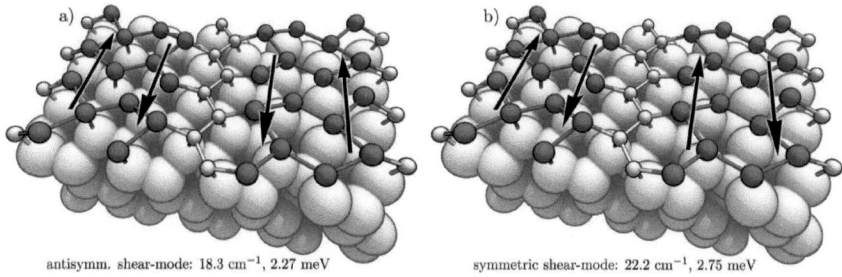

Figure 1.5: *Upon the (4×1)→(8×2) phase transition the soft shear mode splits in two separate modes, one **a)** antisymmetric at 18.3 cm^{-1} and one **b)** symmetric at 22.2 cm^{-1} due to the (8×2) translational symmetry, as indicated by arrows. This is consistent with the observation by Fleischer et al. [46], that the A'' mode measured at 23.5 cm^{-1} possibly consists of 2 separate modes.*

try. Thus the (8×2) reconstruction features many modes that exhibit elongations both within and perpendicular to the $[1\bar{1}0]$ plane simultaneously. Strictly speaking, the A'/A'' notation scheme is no longer applicable. However, due to its convenience it is still widely applied in the literature to the (8×2) phase as well. Thus in the present work, modes whose percentaged elongation is larger/smaller within the $[1\bar{1}0]$ plane than perpendicular to it are still labeled A'/A'' modes, respectively.

Fig. 1.4a/c shows the Γ-point A'/A'' modes of the (8×2) hexagon model, that are localized within the upper two atomic layers. Red circles and blue triangles indicate the modes' degree of localization at the In atoms and within the uppermost atomic layer, respectively. In between Fig. 1.4c shows the modes measured by Raman spectroscopy in Ref. [46]. Error bars represent the measured line widths. A clear assignment of calculated and measured modes is highly difficult due to the multitude of calculated modes, most of which are inactive in experiment. Below 200 cm^{-1} there are especially many A' modes with a high degree of localization at both the In atoms and within the top atomic layer. Only the three lowest energy modes could be assigned clearly, since exactly four calculated modes are present in this energy range, two of them degenerate. The assignment is indicated by vertical green lines. Above 200 cm^{-1} 4 peaks of localization within the uppermost atomic layer are observed, that coincide with 4 measured modes.

The relatively scarce distributed A'' modes allow for a somewhat clearer assignment, since below 100 cm^{-1} only 9 localized modes are obtained. Fleischer *et al.* [46] obtained 5 modes but commented that the three lower energy modes probably consist of

2, 3, and 2 modes, respectively. The calculated modes have been assigned to the measured ones based on this data. The lowest energy A'' mode – which was shown to be the soft shear mode in the previous section – is now slightly redshifted in accordance with the experiment. Due to the (4×1)→(8×2) change of translational symmetry the shear mode splits into two almost degenerate modes. The shear distortions may occur either in antiphase or in phase, as illustrated in Fig. 1.5a/b, respectively. This is consistent with Fleischer's *et al.* observation that the lowest energy A'' mode is probably comprised of two separate modes [46]. The two measured modes at 184 and 262 cm^{-1}, respectively, could not be assigned since no calculated modes are present in this energy range. However, as mentioned previously a clear distinguishment between A' and A'' modes is no longer possible for the (8×2) structure due to the absence of the mirror symmetry. The missing A'' modes are probably simply categorized as A'. Most modes in this energy range feature elongations of almost equal magnitudes both within and perpendicular to the $[1\bar{1}0]$ plane. The highest energy and last A'' mode was assigned to the localization peak at 441 cm^{-1}.

In analogy to the linear decomposition of the LT phase structural models into the phonon eigenvectors of the ideal (4×1) surface, the same process is possible in the reverse case as well. Starting from the phonon eigenvectors of the (8×2) hexagon reconstruction the ideal (4×1) reconstruction in the (8×2) unit cell can be obtained as a linear combination. Fig. 1.6 shows the optimized linear coefficients in analogy to Fig. 1.3. Besides the already known shear, rotary

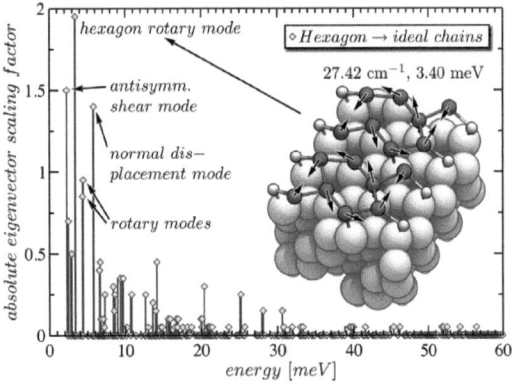

Figure 1.6: *Linear decomposition of the (4×1) ideal reconstruction into the phonon eigenvectors of the (8×2) hexagon structure.*

and normal displacement modes another new mode is present. This mode was termed *hexagon rotary mode* (cf. Fig. 1.6 inset) and also facilitates the reversal of the trimerization similar to the 4-atom rotary modes described in the previous section. It is also marked in the phonon spectra illustrated in Fig. 1.4a. The average displacement per In atom amounts to 0.56 Å between the (8×2) hexagon and (4×1) ideal structure within the (8×2) unit cell. Based on the shown optimized coefficients an average displacement per In atom of 0.04 Å between the linear combination and the (4×1) ideal structure is achieved. If only the modes marked in Fig. 1.6 are taken into account an accuracy of 0.18 Å per In atom is obtained.

Hence these soft phonon modes do not only facilitate the (4×1)→(8×2) phase transition upon cooling the system. The same modes in conjunction with the newly arising hexagon rotary mode also support the reverse (8×2)→(4×1) transition upon heating the nanowire array.

6.2 Temperature-dependent transport properties

The present section aims at deriving the temperature-dependent transport properties for the In/Si(111) nanowire system including the phase transition. On the basis of the results of the previous chapter the In nanowire array's true ground state could be identified. This allows for a combined *frozen-phonon* (FP) and *molecular dynamics* (MD) approach, starting from the hexagon model for the ground state. The FP approach incorporates the temperature from a fully quantum mechanical point of view, but approximates the system's potential by a harmonic ansatz. Hence this approach is well suited for the low temperature regime. On the other hand, MD employs the correct potential, but the energy distribution is treated classically. Thus MD is expected to be accurate at higher temperatures. In principle, combining both approaches should yield accurate results over the entire temperature range.

6.2.1 *Frozen-phonon* (FP) approach

As discussed briefly in chapter 4.2.3 the phonon eigenmodes and -frequencies can be obtained from DFT by calculating the linear force coefficients for each atom in each spatial direction and solving the associated algebraic eigenvalue problem $\omega^2 \mathcal{M} u = \mathcal{K} u$, where $\mathcal{M}$ and $\mathcal{K}$ denote the mass and force constant matrices, respectively. For the elongation u' a harmonic ansatz is assumed

$$u'_i(t) = \frac{1}{\sqrt{M_i}} \sum_m b_m u_{i,m} \cdot e^{-i(\omega_m t + \phi_m)}, \qquad (2.2)$$

where i and m are the atomic and modal indices, respectively. M_i is the mass of atom i, b_m represents a scaling factor for the eigenvector and $u_{i,m}$ denotes the part of the eigenvector u_m of the mode m that belongs to the atom i. To incorporate the temperature T from a quantum mechanical point of view the resulting modes m are occupied according to the Bose-Einstein distribution $\tilde{n}_m(T)$, given by:

$$\tilde{n}_m(T) = \frac{1}{e^{(\hbar\omega_m)/(k_BT)} - 1} \qquad (2.3)$$

Then the energy expectation value of the mode m with the frequency ω_m at the temperature T is $E = \left(\tilde{n}_m(T) + \frac{1}{2}\right)\hbar\omega_m$. By relating the potential energy resulting from the maximum elongation of the isolated mode m to the expectation value of the energy according to

$$\sum_i \frac{1}{2} M_i \omega_m^2 \left(\frac{1}{\sqrt{M_i}} b_m(T) u_{i,m}\right)^2 = \left(\tilde{n}_m(T) + \frac{1}{2}\right)\hbar\omega_m \qquad (2.4)$$

one can obtain the temperature-dependent scaling factor $b_m(T)$. Inserting $b_m(T)$ into Eq. (2.2) allows to determine the temperature-dependent elongation u'_i of any atom i. Subsequently, the phase factor ϕ_m is determined randomly for any mode m. By calculating u'_i and elongating the corresponding atoms accordingly, a sufficiently large number of random configurations is generated for the respective temperature T. The electron transport properties at T are then obtained by calculating the transport properties for each individual configuration and subsequent averaging. As discussed in detail in chapter 2.2.2c the conductance is obtained in linear response according to

$$G = \frac{2e^2}{h} \int dE\, \bar{T}(E) \left(-\frac{\partial f_0}{\partial E}\right), \quad f_0(E) = \frac{1}{e^{(E-E_F)/(k_BT)} + 1} \qquad (2.5)$$

where $\bar{T}(E)$ and f_0 denote the transmission function and the Fermi distribution, respectively. Since linear force constants have been assumed the elongations must be sufficiently small to ensure the validity of this approximation. Thus the temperature may not become too large as well. Due to the quantum mechanical treatment of the energy distribution this approach is expected to be highly accurate at low temperatures.

6.2.2 *Molecular dynamics* (MD) approach

For higher temperatures and larger elongations – especially near and beyond the phase transition's critical temperature – the force constants are not linear anymore. Hence the harmonic approximation assumed in the FP approach is no longer valid. This temperature range is treated by *first principles* molecular dynamics (MD) calculations. The MD approach allows to incorporate the exact quantum mechanical forces. However, the temperature can only be treated classically. Initially the simulation tem-

6.2. Temperature-dependent transport properties

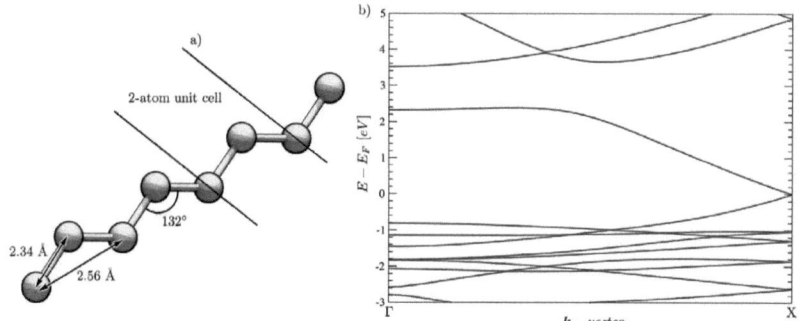

Figure 2.7: *a) Ground state structure of monoatomic Au zigzag chains. b) Band structure for an infinite zigzag Au chain along the chain direction.*

perature T_S is set according to

$$T_S = \frac{1}{f \cdot k_B T} \sum_i M_i \|v_i\|^2, \tag{2.6}$$

where f is the number of degrees of freedom[1]. The velocities are chosen randomly, so that they fulfill the Maxwell-Boltzmann distribution. Treating the system as a canonical ensemble it is equilibrated employing the algorithm of Nosé [194, 195, 196]. After a sufficiently large equilibration time t_{eq} the desired number of configurations can be extracted at randomly chosen MD timesteps for $t > t_{eq}$.

Subsequently, the procedure is analogous to the FP approach: the electron transport properties are obtained individually for each configuration in linear response and are subsequently averaged. The difference between both approaches is only the way the random configurations are obtained. One is accurate for lower temperatures, while the other applies for higher temperatures. In an intermediate temperature range both approaches are expected to yield identical results. This is tested in the following section.

6.2.3 A simple test system: zigzag Au wires

One of the most intensively investigated test systems in electron transport theory are monoatomic Au wires. They can be produced in break junctions between Au tips up to a length of about 8 atoms [7]. Fig. 2.7a shows the zigzag ground state structure of these chains. Close to the Fermi energy these chains exhibit a single metallic band (cf.

---

[1] Here: 3 times the total number of movable atoms within the unit cell.

Fig. 2.7b). Due to their simplicity they represent an ideal test system for the combined FP/MD approach as well.

The following frozen-phonon calculations were performed in a 8-atom unit cell. With respect to the Brillouin zone of the 2-atom unit cell this calculations yields the Γ-point modes as well as the X-point and $\pi/2a$ modes, where a = 4.68 Å denotes the length of the 2-atom unit cell. By means of the elongation patterns the resulting modes were categorized in terms of the six phonon branches: one longitudinal optic and acoustic branch each, and two transverse optic and acoustic branches, respectively.

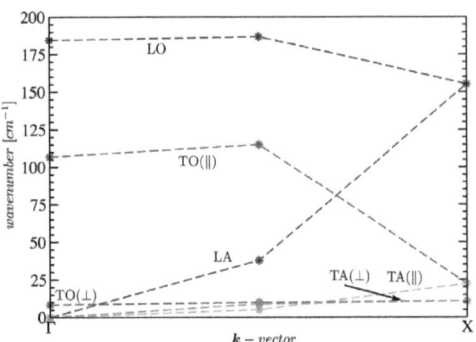

Figure 2.8: *Phonon dispersion relation of the zigzag Au chain. ⊥/∥ denote polarization directions perpendicular and parallel to the plane of the Au chains [197].*

The acoustic Γ-point modes do not meet exactly at the origin due to numerical noise. The resulting dispersion relation shown in Fig. 2.8 has been corrected for these effects. As a result of the high symmetry along the chain direction optic and acoustic branches of equal polarization become degenerate at the X-point. The TO branch lies energetically lower than the two other optic branches, since the coupling is very weak in this direction.

The amplitudes of these phonon modes for Bose-Einstein occupation at 150 K and 300 K, respectively, have been calculated employing the procedure described above. 20 configurations with random phase factors ϕ_m were generated. The same number of random configurations was extracted from MD calculations after an equilibration time of 15.000 time steps with δt = 2 fs. For this conceptual test periodic boundary conditions were assumed for the transport calculations as well. Therefore Eq. (2.5) is reduced to

$$G = \frac{2e^2}{h} \int dE\ N(E) \left(-\frac{\partial f_0}{\partial E}\right), \quad (2.7)$$

where $M(E)$ denotes the number of bands at the energy E. Extracting $M(E)$ from band structure calculations for each one of the random configurations obtained by either FP or MD calculations and evaluating Eq. (2.7) yields the conductances listed

6.2. Temperature-dependent transport properties

	Frozen-Phonon	Molecular Dynamics
150K	0.39 ± 0.17	0.40 ± 0.26
300K	0.44 ± 0.20	0.46 ± 0.25

Table 2.2: *Average conductances and standard deviations in units of $2e^2/h$ in linear response for infinite zigzag Au chains at finite temperatures [197].*

in Tab. 2.2. Both approaches result in conductances that match very closely. The MD approach exhibits slightly higher standard deviations than the FP approach. Interestingly, the average conductances at 150K and 300K are somewhat smaller than the conductance of the equilibrium structure. This can be understood in terms of the symmetry-breaking longitudinal phonon modes. As a consequence of their occupation the degeneracy of the bands at the X-point of the Brillouin zone is lifted, resulting in a small bandgap at the Fermi energy. Since the conductances obtained from both approaches match very closely, the combination of both approaches seems reasonable.

6.2.4 Quasiparticle corrections and eigenstate symmetries

As it turns out this approach cannot be directly applied to the In/Si(111)-(4×1)/(8×2) surface. The (8×2) LT phase is seriously affected by the DFT bandgap problem. In DFT-LDA the gap energy amounts to only $E_{G,LDA} = 0.012$ eV as opposed to the experimental values between 0.1–0.3 eV [30, 36, 42, 44]. As a result the phase transition with respect to the conductance would be obtained at much lower temperatures than T_C. Thus quasiparticle corrections need to be taken into account. However, it is clearly impossible to perform quasiparticle calculations for each configuration. The method of choice is to derive a reasonable scissors operator, that can subsequently be applied to any configuration. Unfortunately, this is not as straight forward as it seems on first glance.

Fig. 2.9a shows the band structure of a distorted In nanowire in the (8×4) unit cell. Since the DFT code can only sort the eigenvalues according to their respective energy, the "band structure" shown is not in actuality a band structure in the strict sense of the word, but rather a loose collection of eigenvalues. Any connections drawn are purely artificial. This is very important to note: from the DFT eigenvalues alone it is impossible to tell which eigenvalues belong to the same band. However, exactly this information is crucial for the application of the scissors operator because the conduction bands are affected much stronger by quasiparticle corrections than the valence bands. The bands shown in Fig. 2.9a are not well separated and exhibit several pos-

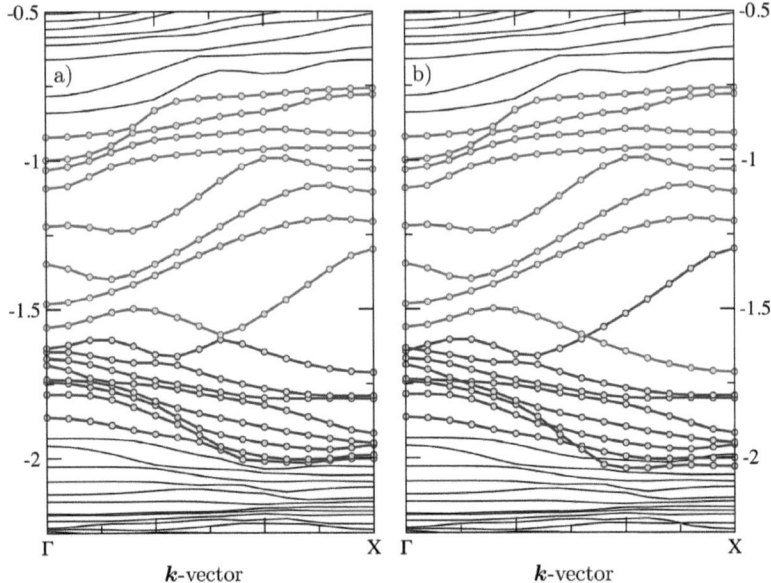

Figure 2.9: *a) DFT-LDA band structure of a random In/Si(111)-(8×2) MD configuration. Conduction, valence and Si bulk bands are drawn in red, blue and black, respectively. b) The same band structure sorted according to the respective eigenstate symmetry, see text.*

sible degeneracies. Therefore it is unknown which eigenvalues represent conduction states and should be shifted.

Knowing which eigenvalues belong to the same band is crucial for another reason as well: to apply the linear response formula from Eq. (2.7) requires to know the number of bands at every energy E with much higher resolution than is numerically feasible by k-point sampling alone. For both reasons it is essential to be able to distinguish crossings from anti-crossings in the band structure and obtain information which eigenvalues together constitute a band. Note that this problem did not arise for the Au chains in the previous section, since only one band was present near the Fermi energy.

To solve this problem an algorithm was developed that assigns the eigenvalues to their respective bands by means of the wavefunction symmetries. In the first step the wavefunctions ϕ_{nk} are projected onto spherical harmonics Y^i_{lm} that are centered at the positions of the ions i, according to:

6.2. Temperature-dependent transport properties

$$P_{nk}^{ilm} = \langle Y_{lm}^{i} | \phi_{nk} \rangle \tag{2.8}$$

If the k-point sampling is not too coarse it can be safely assumed that the symmetry of the wavefunctions belonging to the same bands – and hence the site-specific projections – change little between adjacent k-points. On the other hand, site-specific projections of states belonging to different bands are expected to be notably different as well. At each k-point and band index pair nk the following expression is calculated for any $n < n' < n_{max}$:

$$Q_{nn'k} = \frac{\sum_{i,l,m} P_{nk}^{ilm} \cdot P_{n'k+1}^{ilm}}{\sum_{i,l,m} P_{nk}^{ilm} \cdot P_{nk+1}^{ilm} + \delta x} \tag{2.9}$$

The infinitesimal term δx is introduced in the – unlikely – case that the denominator would otherwise amount to zero. This ratio between the overlaps of the projections becomes $Q_{nn'k} \gg 1$ if a crossing is present between k and $k+1$ and the states ϕ_{nk} and $\phi_{n'k+1}$ belong to the same band.

In practice the implementation is performed by programming 6 nested loops over the indices n, n', k, i, l, m and thus systematically scan the band structure for crossings. The results are stored as "hopping indices" from one band index to another in a $(n \times k)$-matrix. If a crossing is indicated by a $Q_{nn'k} \gg 1$ the corresponding matrix element is set to the difference between the DFT bandindex n and the bandindex of the band to which the eigenstate actually belongs to. If no crossing is present the matrix element is set to zero. Calculating this "hopping" matrix allows to efficiently sort the original DFT eigenvalues according to their respective bands. The described algorithm is found to work very well in case of the In/Si(111) nanowire system.

The original DFT band structure (cf. Fig. 2.9a) could have represented a case similar to Ge, which is erroneously metallic within DFT. While Ge features an indirect gap, the DFT bands are so close in energy that the gap vanishes. Applying this algorithm to the eigenvalues depicted in Fig. 2.9a yields the band structure shown in Fig. 2.9b. Now it is clear that the present case is completely unlike the Ge case and that this band structure is indeed metallic. It will remain so also after introducing quasiparticle corrections by means of a scissors operator. The combined FP/MD approach and this eigenvalue sorting algorithm now enable the treatment of the In/Si(111) nanowire array.

6.2.5 Results of the combined FP and MD approaches

The (8×2) hexagon structural model is used as a starting point for the following calculations, since in the previous chapter it has been confirmed to describe the properties of the In/Si(111)-(8×2) LT phase very well. For this reason DFT-LDA with the In $4d$ states frozen into the core is employed because this is the only approximation which predicts the (8×2) hexagon structure to be stable over the trimer model. The k-space integrations are performed using uniform meshes equivalent to 256 k-points in the (1×1) surface Brillouin zone. For both the FP and MD calculations the surface was simulated by an asymmetric slab with 4 Si bilayers and a vacuum region equivalent to 6 Si bilayers in length. To enable the calculation of Γ- and X-point phonon modes with respect to the (8×2) surface Brillouin zone both the FP and MD calculations have been carried out in the (8×4) unit cell. This is the largest unit cell that was still numerically feasible on the available systems[2].

In all cases a number of 30 random configurations has been employed at 50 K and 300 K. Using the FP approach the configurations were obtained for Bose-Einstein occupation of the phonon modes with random phase factors, as described in section 6.2.1. To estimate the influence of zero-point vibrations another 30 random configurations were generated for zero-point occupation only. The MD configurations were extracted randomly after an equilibration time of 7700 steps with a step size of $\delta t = 2$ fs. For each of the random configurations obtained by either the FP or MD approach the band structure was calculated in ΓX-direction employing a $16 \times 1 \times 1$ k-point grid. Afterwards, the DFT eigenvalues were assigned to their respective bands by means of the eigenstate symmetry algorithm discussed above. A constant scissors shift of 0.5 eV is applied to the identified conduction states that lie more than 0.5 eV above the highest occupied electronic state, ensuring an approximate reproduction of the measured Si bulk band structure. For the energetically lower lying conduction states a linear interpolation up to a zero shift for states directly at the Fermi energy is used in order to model the energy-dependence of the electronic self-energy. Subsequently, the conductance can be obtained in linear response from the resulting band structures.

The averaged conductances and displacements of the In atoms from their respective equilibrium positions with regard to the hexagon model are compiled in Tab. 2.3. At $T = 0$ K the average displacement amounts to 0.08 Å, which is well within the range of validity for the harmonic approximation. The conductance in linear response is zero.

---

[2] Both the FP and MD calculations ran for several months on a fast NEC-SX9 vector supercomputer.

6.3. Discussion

	FROZEN-PHONON		MOLECULAR DYNAMICS	
	G [$2e^2/h$]	Δ [Å]	G [$2e^2/h$]	$\bar{\Delta}$ [Å]
0 K	0.00 ± 0.00	0.08 ± 0.03	-	-
50 K	0.11 ± 0.28	0.12 ± 0.05	0.22 ± 0.36	0.18 ± 0.08
300 K	0.50 ± 0.32	0.27 ± 0.11	1.95 ± 0.86	0.65 ± 0.26

Table 2.3: *Average conductances in linear response and average displacements of the In atoms from their equilibrium positions within the Si(111)-(8×4)In hexagon model.*

Hence zero-point vibrations do not affect the conductance. Increasing the temperature to $T = 50$ K yields a small increase in conductivity to 0.11 and 0.22 $2e^2/h$ for both the FP and MD approaches, respectively. However, the MD value is already twice as large as obtained within the frozen-phonon approximation. This behaviour occurs mainly due to the classical treatment of the temperature by MD since the modal occupation does not correspond to the Bose-Einstein distribution. As the average displacement amounts to 0.18 Å ±0.08 Å anharmonic effects are beginning to arise as well. Still, 0.12 Å as opposed to 0.18 Å obtained by the FP and MD approaches, respectively, is a rather reasonable agreement between both approaches.

At 300 K a conductance of 0.5 $2e^2/h$ is obtained in frozen-phonon approximation. As very few of the FP configurations are metallic this increase in conductivity can be attributed mainly to thermal smearing of the Fermi distribution (cf. Eq. (2.7)). On the other hand, the MD approach yields a much higher conductivity of 1.95 $2e^2/h$ for the two chains in the (8×2) unit cell. However, the experimental value of G = 200 μS per chain [33, 44] is significantly higher than even the MD value of $G = 1/2 \cdot 1.95\, 2e^2/h = 75.5\, \mu$S.

6.3 Discussion

Obviously, within the frozen-phonon approximation no insulator-metal transition is obtained. Only a thermal smearing is observed. Fig. 3.10c shows the radial distribution of the In atoms at 50 K and 300 K. Due to the harmonic approximation the mean position is always the (8×2) hexagon model, as indicated by white stars. The radial distributions are highly localized and never reach the ideal (4×1) reconstruction. Thus the harmonic potential implied by the frozen-phonon approximation effectively prohibits a modeling of the phase transition. However, for temperatures sufficiently below the critical temperature T_C the calculated conductivity within the FP approach is correct.

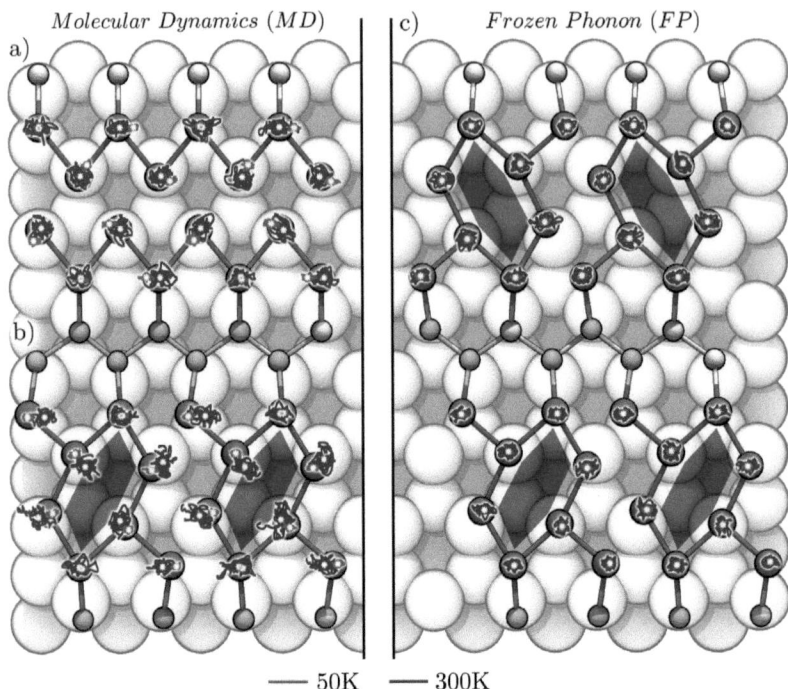

Figure 3.10: *Radial distribution of the In atoms obtained from the **a/b**) MD and **c**) FP approaches. The mean positions are indicated by white stars. At 50 K the mean MD positions correspond to the hexagon model (**a**), while at 300 K the In atoms oscillate between the two degenerate hexagon configurations and assume the ideal (4×1) reconstruction as the mean configuration (**b**).*

On the other hand the MD simulations exhibit a very different behaviour. At 50 K the mean configuration remains the (8×2) hexagon model (cf. Fig. 3.10b). Increasing the temperature to 300 K yields the ideal chains in (4×1) reconstruction as the mean configuration. However, the radial distribution in Fig. 3.10a/b shows that the chains oscillate between the two degenerate (8×2) hexagon structures with the ideal (4×1) reconstruction as an intermediate structure. This behaviour contradicts the experimental findings and explains the underestimation of the experimentally determined conductivity. A photoemission experiment by Yeom *et al.* [39] reveals a sudden change from one well defined band structure to another at the transition temperature T_C. Since these experiments take place on a much faster time scale than the oscillations predicted by the MD simulations, the presence of such fluctuations would be clearly visible. Fleischer *et al.* [46] also observe well defined LT and RT

6.3. Discussion

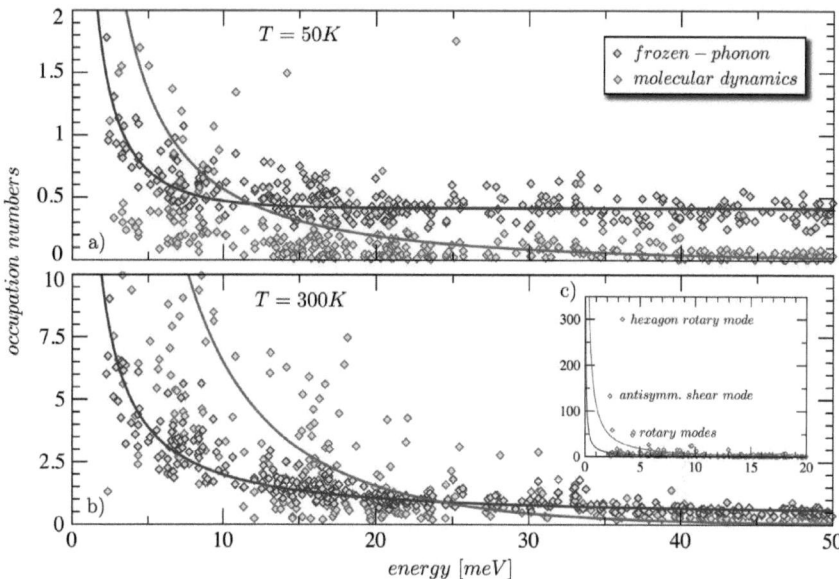

Figure 3.11: *Linear combination of each of the 30 MD and FP configurations, respectively, in terms of the frozen-phonon eigenvectors. The resulting eigenvector scaling factors have been transformed into occupation numbers assuming a harmonic potential according to Eq. (2.4). The inset **(c)** shows a larger section of diagram **(b)**, see text. The solid lines are filled according to $n(E) = a/(exp(E/(k_B T)) - 1) + b$, with a, b and T as fitting parameters. For the FP configurations only T and b are fitted, effectively retaining the Bose-Einstein distribution including zero point occupation.*

phonon spectra, while these oscillations imply that at least some of the LT phonon modes would be visible also at room temperature. Hence these MD oscillations are expected to be unphysical. Assuming that the conductivity measurements in Ref. [33, 44] represent the conductivity of stable and well defined LT and RT structures, it is clear that the calculated conductance of a chain oscillating between the RT and LT phases significantly underestimates the measured conductance at room temperature. This erroneous behaviour of the MD is most probably related to its ignorance of the quantum mechanical phonon mode occupations.

To allow for a closer comparison between the FP and MD approaches each of the 30 respective FP and MD configurations at 50 K and 300 K has been factorized into a linear combination of the frozen-phonon eigenvectors of the (8×2) hexagon structure. This is still possible even for the 300 K MD configurations. All of the linear combinations achieve an accuracy of at least 0.01 Å average displacement of the In atoms with respect to the original structure. Assuming a harmonic potential the resulting eigenvector scaling factors can be transformed into phonon mode occupation numbers according to Eq. (2.4). These "occupations" for both the MD and FP configurations are shown in Fig. 3.11. Naturally, in case of the FP configurations the Bose-Einstein distribution is retrieved within the numerical accuracy (cf. blue lines, Fig. 3.11), since the FP configurations have been originally generated employing the Bose-Einstein distribution. At 50 K the factorization of the MD configurations into phonon modes numbers yields occupation numbers of roughly the same magnitude. Some modes feature slightly higher occupation numbers than is the case for the FP configurations. Thus anharmonicities of the potential are not yet pronounced but noteworthy. The spread of the occupation numbers is significantly higher though and the Bose-Einstein distribution yields only a fit with poor correlation. Instead the function

$$n(E) = \frac{a}{e^{E/(k_B T)} - 1} \tag{3.10}$$

was employed with a, T as fitting parameters (cf. red lines, Fig. 3.11). While the MD correctly takes anharmonicities into account it does not reflect the Bose-Einstein occupation of the phonon modes due to its classical treatment of the temperature. Zero-point vibrations are missing as well, which is seen in the phonon spectra especially towards higher energies.

At 300 K the spread of the occupation numbers for the MD configurations as well as the distance between the distributions increase even more. Still most occupation numbers are of the same order of magnitude as the FP occupations (cf. Fig. 3.11b). However, there are five modes that feature large elongations that are completely beyond the harmonic approximation, as shown in the inset in Fig. 3.11c. These are exactly the modes that transform the (8×2) hexagon structure into the (4×1) ideal reconstruction. Thus the (4×1) phase can never be reached by Bose-Einstein occupation of the (8×2) hexagon structure's phonon modes in frozen-phonon approximation. However, while the harmonic approximation prohibits the modeling of the phase transition, the frozen-phonon eigenvectors correctly span the phase space of the system. Thus employing the frozen-phonon eigenvectors in combination with the correct potential should enable an accurate treatment of the phase transition.

In conclusion, the derivation of the temperature dependent conductivity is correct only for temperatures significantly below the critical temperature T_C. The phase transition cannot be modeled using the frozen-phonon approximation. MD calculations

6.3. Discussion

in contrast do exhibit an insulator-metal transition, but predict unphysical oscillations between the two degenerate (8×2) hexagon structures at room temperature. Thus a classical treatment of the energy distribution is too inaccurate for this highly subtle system. Instead the phase transition must be treated from a fully quantum mechanical point of view including entropy contributions. The prospect of a suitable method is discussed further in the outlook.

Furthermore, the frozen-phonon calculations largely explain the experimental Raman spectra. Several soft phonon modes have been identified that facilitate the phase transition, providing a strong indication that the phase transition is indeed phonon mode driven. These results have yet to be published.

A journey of a thousand miles begins with the first step.

– Lao Tse

Chapter 7
Summary and conclusions

Since the invention of the electric tabulating machine by Hollerith in 1889, computing devices have been consistently increased in speed and reduced in size at the same time. This process, as governed by Moore's famous law, represents a major economic factor for the electronics and semiconductor industry. However, today's microelectronics faces a very difficult challenge: as devices are beginning to approach the atomistic scale, quantum effects will become visible at room temperature. Both these urgent technological needs and interests in fundamental physics are currently driving a huge research effort into the properties of atomic-scale nanowires. Despite the importance of atomic-size nanowires for future nanoelectronic applications few such systems are currently known and the known ones are barely understood today. One of the most intensively studied structures is the prototypical In/Si(111)-(4×1) nanowire array. Ever since the discovery of its quasi-1D metallic nature and associated (4×1)→(8×2) phase transition more than 10 years ago, this system has been intensively and controversially discussed.

The present work treats a heterogeneous ensemble of problems related to the understanding and theoretical modeling of these nanowires by large-scale *first principles* computer simulations. With respect to future nanoelectronic device concepts this work aims at generating a solid knowledge and detailed understanding about the physical foundations of such perspective future devices. The prototypical In/Si(111)-(4×1)/(8×2) nanowire array is employed as a suitable model system.

7.1 Results of the present work

It was demonstrated that the intensive and ongoing search for the internal structure of the low temperature (LT) phase of the In/Si(111)-(4×1)/(8×2) nanowire array cannot be concluded by the surface energetics alone. Based solely on the surface energies, present state-of-the-art *first principles* calculations are not accurate enough to allow for an unambiguous identification of the LT phase geometry. Several different structural models have been examined: the *trimer* and *hexagon* models in (4×2) and (8×2) translational symmetry.

While the relative stability of these models depends on the exact numerical details, i. e. the treatment of exchange and correlation effects and the In 4d states, the calculated electronic structure and transport properties, however, strongly indicate the formation of hexagons. In contrast to In atom pairing and trimerization, hexagon formation opens a fundamental energy gap.

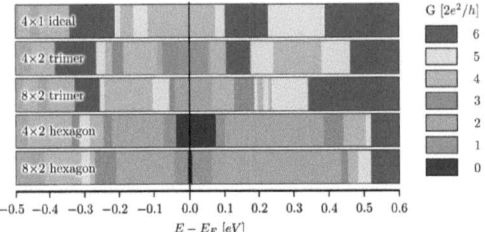

Figure 1.1: *Quantum conductance spectrum for electron transport along the chain direction calculated for In/Si(111) model structures.*

The quantum conductance of infinite In nanowires that model the surface chain structures explains the change of the measured wire resistance for variations between the room temperature (RT) and LT phases of the In/Si(111) system, provided hexagons form (cf. Fig. 1.1).

These results demonstrate the distinct influence of small changes of the nanowire geometry on its conductance. Given the extremely flat potential energy surface, it is not difficult to envisage "1D devices" where the electron transport can be tuned at will (cf. Publ. [13,14]).

From a technological point of view the effect of impurities and defects on the electron transport properties is highly interesting, as current microelectronics is based almost entirely on the concept of tuning and modulating electronic device characteristics by the controlled creation of defects or impurity doping. The In nanowire array is a very interesting test bed also for conductance modification at the atomic scale: it is well accessible to both experiment and *first principles* theory and thus helps to gain a deeper understanding of nanoscale electron transport. *First principles* calculations performed for ideal and adatom-deposited In nanowires predict an adatom-specific, and in some cases very pronounced, decrease of the wire conductivity upon adatom deposition. For In adatoms – where measurements exist – the reduction by more than one third agrees with the existing data. The adsorption of hydrogen does not substantially reduce the conductance, which is also in agreement with existing measurements. For O

7.1. Results of the present work

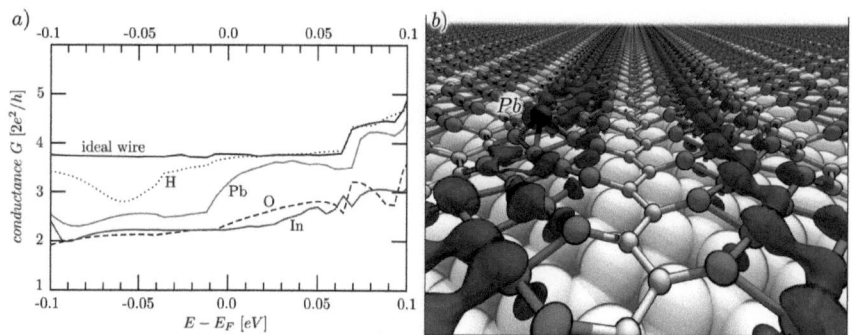

Figure 1.2: *a) Quantum conductance spectra for electron transport along the wire direction calculated for ideal and adatom-modified In/Si(111) structures. b) Isodensity surface of the local DOS at E_F, illustrating the local DOS at the ideal In nanowires and its modification upon Pb adsorption.*

deposition, the calculations predict a similar drop in conductance, whereas the impact of Pb adatoms is slightly smaller (cf. Fig. 1.2a).

The nanowire conductance modification due to the adatoms can be traced to different mechanisms: potential-well scattering (Pb), nanowire deformation (In), or a combination of both effects (O) (cf. Publ. [11]). An interpretation of the conductance drop in terms of perturbed conduction channels may serve as an illustrative example (cf. Fig. 1.2b). Possible direct applications of nanowire conductance modification effects include sensing and biosensing devices. I. e. the detection of viruses and tumor cells has recently been demonstrated by conductance modification effects of antibody-adsorbed carbon nanotubes [201, 202].

One of the most intensively studied and controversely discussed properties of the In/Si(111) nanowire array is the internal structure of its associated LT (8×2) ground state. Given the ambiguities of the total-energy calculations in determining the ground state structure of the (8×2) phase, the comparison of optical fingerprints calculated for structural candidates with measured data is expected to be very helpful. The optical reflectance anisotropies (RA) for ideal and hexagon as well as trimer reconstructed models for the In/Si(111) nanowire array have been calculated from *first principles*. Within the visible spectral range the optical signatures of both the trimer and hexagon models are very well suitable to explain the spectral changes aquired during the formation of the LT phase. A prediction of the reconstruction model is not possible from the visible spectral range alone. However, the small differences in geometry between the competing structural models lead to significant changes in the band structure near the Fermi level. Thus the extension of the optical measurements

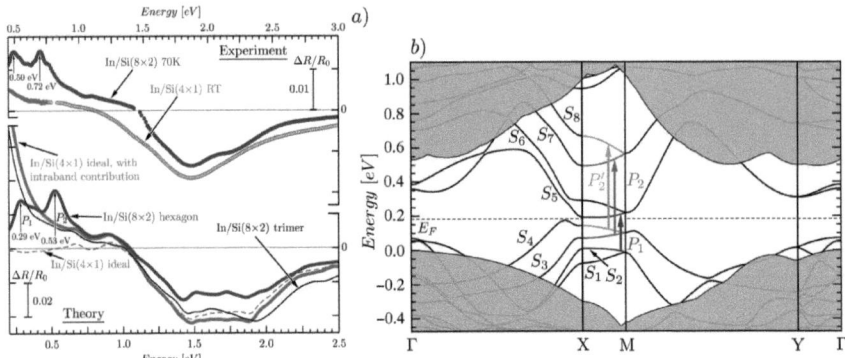

Figure 1.3: *a) Calculated and measured RA-spectra for In/Si(111)-(4×1)/(8×2). b) Band structure and pronounced anisotropic optical transitions of the hexagon model for the In/Si(111)-(8×2) surface.*

and calculations into the mid-infrared regime is expected to be more conclusive. *First principles* calculations of the mid-infrared reflectance anisotropy as well as the intraband contributions have been performed for the In/Si(111) nanowire array for the first time. Two strong anisotropic optical peaks in the mid-infrared regime were predicted for the hexagon model. In contrast, a steep rise of the RA-spectra in analogy to a Drude-like behaviour was obtained for the trimer model. Recent measurements indeed observe the formation of these two anisotropic peaks without any indication for a remaining Drude-tail. In comparison with this data the trimer model can be excluded (cf. Fig. 1.3).

The close agreement between measured data and spectra calculated for the hexagon model provide very strong evidence in favor of the hexagon model for the In nanowire ground state. These results effectively conclude the search that has been ongoing for more than 10 years (cf. Publ. [3,8,9]).

While the hexagon model could thus be confirmed as the long-sought internal structure of the LT ground state, the question regarding the driving force of the phase transition is still open. Both the In/Si(111)-(4×1)/(8×2) surface's thermal properties as well as the phase transition itself have been studied previously by Raman spectroscopy [46] and temperature-dependent transport measurements. To achieve a detailed theoretical understanding of these data the calculation of phonon modes and large-scale molecular dynamics calculations have been performed. The calculated phonon spectra are suitable to explain both the RT and LT Raman spectroscopy mea-

7.1. Results of the present work

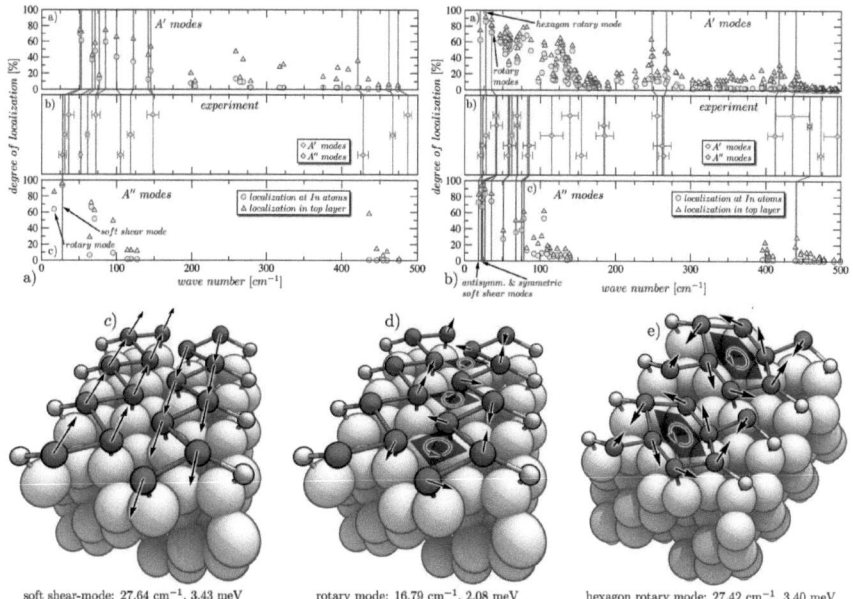

Figure 1.4: *Comparison of theoretical and experimental* **a)** *HT and* **b)** *LT phonon spectra. Low energy* **c)** *shear,* **d)** *rotary and* **e)** *hexagon rotary modes, that facilitate* $(4\times1)\leftrightarrow(8\times2)$ *the phase transition.*

surements (cf. Fig. 1.4a/b). González et al. suggested a soft shear mode as a possible driving force of the phase transition [35]. By comparing the experimental data to the present work's phonon calculations the existence of such a soft shear mode could indeed be confirmed for the first time. At the same time it was demonstrated that the soft shear mode represents only a part of the total picture: two more distinct soft modes could be identified that facilitate the transition between the high and low temperature phases (cf. Fig. 1.4c-d). These rotary and hexagon rotary modes are suitable to explain the three lowest energy A' modes observed in the measured Raman spectra of the LT phase.

In conjunction with the observed soft shear mode these three soft modes are expected to facilitate the $(4\times1)\leftrightarrow(8\times2)$ *phase transition.*

To explain the temperature-dependent transport measurements a method was developed to combine frozen-phonon calculations and molecular dynamics simulations in order to calculate the nanowires' conductance in linear response over the entire tem-

Chapter 7. Summary and conclusions

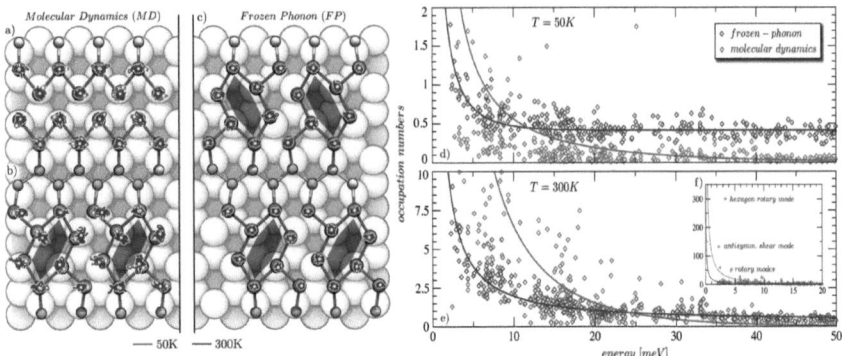

Figure 1.5: *Radial distribution of the In atoms obtained from the* **a/b)** *MD and* **c)** *FP approaches. The mean positions are indicated by white stars.* **d/e)** *Linear combination of each of the 30 MD and FP configurations, respectively, in terms of harmonic occupations of the frozen-phonon eigenvectors (cf. Eq. 2.4). The inset* **(f)** *shows a larger section of diagram* **(e)**. *The solid lines are filled according to* $n(E) = a/(\exp(E/(k_B T)) - 1) + b$, *with a, b and T as fitting parameters.*

perature range. While the phonon and molecular dynamics calculations are suitable for low and high temperatures, respectively, it was demonstrated that neither method is able to correctly address the temperature range near the critical temperature T_C of the phase transition. The harmonic approximation inherent to the frozen-phonon approach effectively prohibits any modeling of the phase transition. On the other hand, the molecular dynamics simulations suffer from the classical treatment of the energy distribution, resulting in unphysical oscillations between the two degenerate hexagon ground states at room temperature (cf. Fig. 1.5).

It turns out that the mechanisms involved in the phase transition of this highly subtle system are purely quantum mechanical in nature and can only be treated from a fully quantum mechanical point of view. A suitable approach is presented in the outlook. These results have yet to be published.

The present work as a whole contributes to the understanding of structural, electron transport, optical and thermal properties of nanoscale systems as encountered upon further size reduction of current microelectronics. It also represents an important building block towards doping studies including electron donor and acceptor effects with respect to future nanowire device applications, as discussed in the outlook.

Furthermore, the present author has also pursued other fields of study, such as optical excitation spectra in many-body perturbation theory at the example of LiNbO$_3$ (cf. Publ. [12]) and the interaction of water with solid surfaces (cf. Publ. [10]), partially in collaboration with the local chemistry group of Prof. G. Grundmeier (cf. Publ. [4,5]).

7.2 Outlook

7.2.1 Quantum mechanical treatment of the phase transition

The ability to accurately predict a materials properties based on *first principles* density functional theory (DFT) calculations is a major achievement of modern solid state theory. However, in the present work it was demonstrated that standard DFT approaches have difficulties to describe the In/Si(111)-(4×1)/(8×2) nanowire array's phase transition due to inherent approximations. In order to obtain an accurate description and in-depth understanding of this important system's phase transition several major obstacles have yet to be overcome. Frozen-phonon calculations incorporate the correct quantum mechanical energy distribution, but have to approximate the potential energy landscape. Otherwise the phonon mode calculations for such large systems would be numerically unfeasible. On the other hand, molecular dynamics simulations may employ the exact potential energy landscape, but are restricted by definition to only a classical treatment of the energy distribution. Either approach alone is unable to capture the In/Si(111)-(4×1)/(8×2) surface's intriguing and highly subtle properties for temperatures near the phase transition's critical temperature T_C. Unfortunately, combining both approaches as demonstrated in the present work is still not accurate enough. However, the present work represents an ideal starting point for treating the phase transition from a fully quantum mechanical point of view. As observed in chapter 6, the frozen-phonon calculations do not employ the correct potential, but still span the entire relevant phase space of the system. Starting from the hexagon model for the low temperature ground state even the other degenerate hexagon model could be represented in terms of a linear combination of the frozen-phonon eigenvectors. Any of the MD configurations at 300K could be linearly combined as well. Based on this observation the phase transition is proposed to be modeled by a *quantum Monte Carlo approach* employing the frozen-phonon eigenvectors in combination with the exact potential energy landscape. The calculational scheme works as follows:

1. A random mode of frequency ω_{ki} is selected and randomly elongated in a way that reasonably covers the system's phase space.

2. Subsequently a total energy calculation is performed for the distorted structure.

3. The selected mode's occupation for a given temperature T is determined by the Bose-Einstein distribution. Thus the upper energy limit E is given by:

$$E = \left(\bar{n}_m(\hbar\omega_{ki}) + \frac{1}{2}\right)\hbar\omega_{ki}, \text{ with } \bar{n}_m(\hbar\omega_{ki}) = \frac{1}{e^{\frac{\hbar\omega_{ki}}{k_B T}} - 1} \qquad (2.1)$$

If the resulting energy difference ΔE with respect to zero elongation for the selected mode is below the threshold E the new configuration is adopted. Otherwise it is discarded.

4. Afterwards the cycle begins again with step 1.

The major point is to ensure an efficient coverage of the system's phase space. A sensible approach seems to elongate the modes incrementally in the same direction until ΔE increases beyond E. Then the direction is reversed. The step size could be made dependent on the slope of the respective mode's potential and probably on its occupation as well. In principle this approach should allow for a much more accurate treatment of the phase transition with respect to MD simulations. However, the computational impact is huge permitting only small unit cells. An enormous speedup could be obtained though by taking into account the full potential of only those modes that exhibit strong anharmonicities. In chapter 6 (cf. Fig. 3.11) it was demonstrated that many modes of the In/Si(111)-(8×2) surface towards higher energies feature very similar elongations according to both the frozen-phonon and MD approaches. Employing the harmonic approximation for these modes is expected to enable the treatment of sufficiently large unit cells. This work is in progress.

7.2.2 Entropy contributions

Even zero-temperature conditions are not correctly described by ground-state DFT total-energy calculations, due to the neglect of zero-point contributions. The discrepancies get larger with increasing temperature. To accurately describe the phase stability at finite temperatures all of the respective system's excitations and degrees of freedom involved in creating entropy have to be determined. Generally, phonon contributions and electron entropy are considered to be only small quantitative corrections. However, in cases where different phases are otherwise very close in energy these contributions may become important. The electronic and vibronic parts of the

7.2. Outlook

free energy F are given by [198, 199]

$$F_{el} = -k_B T \int_{-\infty}^{+\infty} dE \, n_F [f \ln f + (1-f)\ln(1-f)] \quad (2.2)$$

$$F_{vib} = \underbrace{\sum_{k,\mu} \frac{1}{2}\hbar\omega_{k,\mu}}_{E_{zero-point}} + \underbrace{k_B T \sum_{k,\mu} \ln\left(1 - e^{-\frac{\hbar\omega_{k,\mu}}{k_B T}}\right)}_{S_{vib}}, \quad (2.3)$$

where n_F, f denote the Fermi distribution and density of states, respectively. $\omega_{k,\mu}$ is the μ-th phonon mode at k-point k. Thus F_{vib} is obtained directly from the phonon spectra. F_{el} can be calculated from the band structure. The Fermi level is determined by the condition [199]

$$\int_{-\infty}^{+\infty} d\epsilon \, n_F(\epsilon) f(\epsilon) = N_{tot}, \quad (2.4)$$

where N_{tot} represents the total number of valence electrons. It would be highly interesting to see whether entropy contributions are important with respect to the phase stability of the different structural models.

In conjunction with the quantum Monte Carlo approach described in the previous section the suggested course of action is expected to allow for an in-depths understanding of the phase transition's driving mechanism. Calculating the entropy contributions is currently in progress.

7.2.3 Doping vs. optical pumping

Nanowire doping has recently become an upcoming "hot topic" that receives strongly increasing attention. With respect to future device applications it would be highly interesting to extend the present work's doping study towards electron donor/acceptor effects and their impact on the nanowires' structural and electronic properties.

In 2009 a tuning of the critical temperature T_C of the In/Si(111)-(4×1)/(8×2) surface's phase transition has been observed experimentally by adatom deposition [56, 64]. Tiny amounts of Na adatoms decreased T_C. This decrease was attributed to the charge transfer from the Na atoms towards the surface. The resulting upwards shift of the Fermi level increases the deviation of the Fermi nesting vector from its ideal condition for a commensurate charge density wave (CDW) transition. Surprisingly, an increase of T_C was observed upon oxygen deposition. As oxygen defects deprive the surface of electrons the Fermi level is lowered, approaching the ideal Fermi-surface nesting

Chapter 7. Summary and conclusions

Figure 2.6: *Low energy electron diffraction (LEED) pictures for **a)** clean and **b)** oxygen-adsorbed In/Si(111) surfaces. Red arrows indicate (8×2) patterns [64]. STM images of In nanowires under **c)** dark and **d)** illuminated conditions, demonstrating an optically induced triggering of the metal-insulator transition [57].*

condition. Thus electron/hole doping affects the In/Si(111) nanowire array in analogy to thermal excitations: by supplying additional electrons or holes either the (4×1) or (8×2) phase is stabilized, respectively.

In 2008 Terada *et al.* demonstrated an optically induced control of the In/Si(111) metal-insulator transition [57]. By tuning the band filling of the In surface states employing optical pumping the nanowires' phase state could be switched back and forth between the (4×1) and (8×2) reconstruction, even at $T > T_C$. Simultaneous conductance measurements confirmed the (4×1) and induced (8×2) reconstructions to be metallic and semiconducting, respectively. Thus Terada *et al.* effectively demonstrated the operation of an atomic-size optical switch, with electron transport occurring solely in the ballistic regime without scattering. Intriguingly, increasing the band filling by photoexcitation was observed to stabilize the (8×2) phase. This is highly unexpected because thus optical pumping affects the nanowires in the reverse way with respect to thermal excitations or impurity doping.

Both optical pumping and impurity doping offer the important prospect of modulating and controlling the nanowires' electrical characteristics in a way that may be very useful in future one-dimensional device applications. However, no satisfying explanation for these observations has been found as of today. An extensive theoretical study of the mechanisms involved is therefore highly suggested. Performing DFT calculations for the In/Si(111)-(4×1)/(8×2) surface with modified electron occupations is expected to be a useful starting point. The present work represents an ideal platform to delve into this fascinating and important field of study.

Chapter 8
Addendum

Shortly after submitting the present dissertation it became apparent that entropy contributions in conjunction with a triple-band Peierls instability are indeed responsible for the phase transition. Thus the origin and nature of the In/Si(111)-(4×1)/(8×2) surface's phase transition are finally revealed. This work is now included as an addendum to the original dissertation and is available online as Publ. [1].

8.1 Entropy explains the phase transition

For a fixed stoichiometry, the ground state of the surface-supported nanowires is characterized by the minimum of the free energy F as a function of the substrate crystal volume V and the temperature T that can be obtained using atomistic thermodynamics, see, e.g. Ref. [203]. Within the adiabatic approximation, F is given by

$$F(V,T) = F_{el}(V,T) + F_{vib}(V,T), \qquad (1.1)$$

with $F_{el} = E_{tot} - TS_{el}$, where the total energy E_{tot} is approximated by the zero-temperature DFT value and the electronic entropy S_{el} is calculated from

$$S_{el} = k_B \int dE\, n_F [f \ln f + (1-f)\ln(1-f)]. \qquad (1.2)$$

Here n_F and f denote the density of electronic states and the Fermi distribution function, respectively. The vibrational free energy of the supercell with volume Ω is calculated in harmonic approximation

$$F_{vib} = \frac{\Omega}{8\pi^3} \int d^3k \sum_i (\frac{1}{2}\hbar\omega_i(\mathbf{k}) + k_B T \ln(1 - e^{-\frac{\hbar\omega_i(\mathbf{k})}{k_B T}})). \qquad (1.3)$$

Theory ω_0 [cm^{-1}]			Experiment ω_0 [cm^{-1}]		
(4×1)	→	(8×2)	(4×1)	→	(8×2)
22	→	20	31 ± 1	→	21 ± 1.6
		27			28 ± 1.3
		hexagon rotary mode			
44	→	47	36 ± 2	→	41 ± 2
51	→	53	52 ± 0.6	→	57 ± 0.7
62	→	58, 69	61 ± 1.3	→	62, 69 ± 1.5
65, 68	→	70, 69, 78, 82	2·72 ± 3.3	→	83 ± 2.3
100, 104	→	97, 106, 113, 142	105 ± 1	→	100–130
129, 131	→	137, 142	118 ± 1	→	139 ± 1.2
143, 145	→	139, 145, 146, 147	2·148 ± 7	→	139, 2·154±2
28	→	18, 19	28 ± 0.9	→	2·23.5 ± 0.8
shear mode					
	→ antisym./sym. shear mode				
		35			3·42 ± 3.5
		51			2·59 ± 3
		75			69 ± 1.5
		82			85 ± 1.7

Table 1.1: *Calculated Γ-point frequencies for strongly surface localized A'/A" (upper/lower part) phonon modes of the Si(111)-(4×1)/(8×2)-In phases in comparison with experimental data [46]. The symmetry assignment of the (8×2) modes is only approximate, due to the reduced surface symmetry.*

The wave-vector dependent phonon frequencies $\omega_i(\mathbf{k})$, as well as the corresponding eigenvectors are obtained from the force constant matrix calculated by assuming $F_{el}(V,T) \sim E_{tot}(T=0)$, i.e., neglecting the explicit temperature and volume dependence. Given the extremely flat potential-energy surface of the Si(111)-In system, that renders already the search for the minimum energy structure a difficult problem [29, 38], Publ. [14], the uncertainty induced by the above approximations is expected to be small compared to the overall error bar of the calculations.

The calculated Γ-point frequencies for strongly surface-localized vibrational modes of the Si(111)-In nanowire array are compiled in Tab. 1.1. The table contains the present results for the (4×1) phase as well as their assignment to the frequencies of geometrically similar eigenvectors of the (8×2) phase in comparison with the Raman data from Fleischer et al. [46]. While measured and calculated frequencies of the (8×2) phase agree typically within a few cm^{-1}, the deviations between experiment and theory are frequently larger for the (4×1) modes. This indicates that anharmonicity effects neglected here affect the RT modes more noticeably than the LT data.

8.1. Entropy explains the phase transition

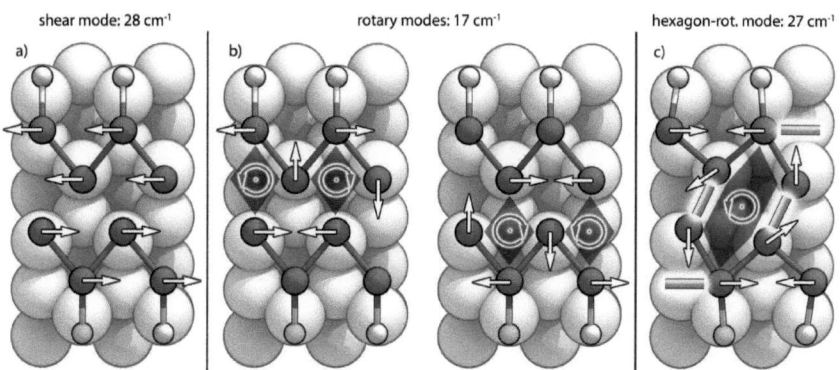

Figure 1.1: *Calculated eigenvectors for three prominent phonons modes (notation as in Tab. 1.1) of the Si(111)-(4×1)-In (a,b) and Si(111)-(8×2)-In phase (c). Golden bars indicate newly formed bonds upon the (4×1)→(8×2) phase transition. The mode shown in (b) – occuring at the X point of the (4×1) BZ – is twofold degenerate due to the existence of an equivalent mode at the neighboring In chain.*

Interestingly, the calculations confirm the existence of a low-frequency shear mode of A″ symmetry for the Si(111)-(4×1)-In phase at 28 cm^{-1}. This mode, which was also detected by Raman spectroscopy [46], is energetically below the phase transition temperature of about $k_B T \sim 83$ cm^{-1} and has been suggested to correspond to the lattice deformation characteristic for the (4×1) ⟶ (8×2) phase transition [35, 37, 204]. The calculated eigenvector of this mode (Fig. 1.1a) shows the two In atom zigzag chains oscillating against each other. This deviates somewhat from earlier predictions [47], but agrees with recent *first-principles* molecular dynamics simulations [35, 37].

In fact, the structural transformation from the In zigzag-chain structure with (4×1) symmetry to the In hexagons with (8×2) translational symmetry can be perfectly described by superimposing the calculated eigenvector of the 28 cm^{-1} mode with the two degenerate low-frequency X-point modes at 17 cm^{-1} (cf. Fig. 1.1a/b). Similarly, the combination of the corresponding shear mode of the Si(111)-(8×2)-In phase at 18 cm^{-1} [1] with the hexagon rotary mode at 27 cm^{-1} (Fig. 1.1c) transforms the In hexagons back to parallel zigzag chains. The calculated phonon modes support the geometrical path for the phase transition proposed in Refs. [35, 37, 204]. In fact, they give an atomistic interpretation of the triple-band Peierls model [34, 36, 204]: The soft

---

[1] The calculated blueshift of the shear mode upon (8×2) ⟶ (4×1) phase transition agrees with the measured frequency shift [46] and provides a strong argument against the dynamical fluctuation model [35, 37, 38]: If at RT the system would fluctuate strongly, thereby frequently visiting configurations associated with the (8×2) structures, the associated phonon mode would redshift rather than blueshift.

shear mode lifts one metallic band above the Fermi energy, while the rotary modes lead to a band-gap opening for the remaining two metallic In surface bands.

However, what exactly is causing the phase transition? Before discussing the difference in the free energies calculated for the phases of the Si(111)-In surface (cf. Fig. 1.2), a word of caution is in order. As pointed out above, the weak corrugation of the In atom potential energy surface leading to small and error-prone force constants as well as the harmonic approximation impair the accuracy of the calculated phonon frequencies. In order to minimize systematical errors, only results obtained for supercells of identical size and numerical parameters are compared.

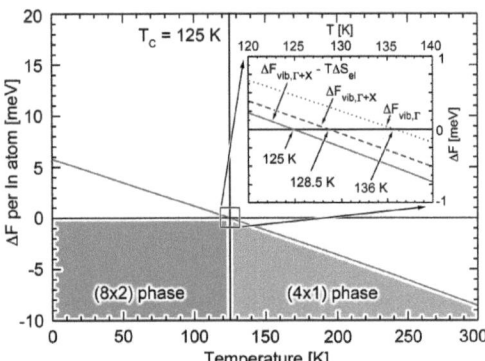

Figure 1.2: Difference of the free energy $F(T)$ calculated for the (4×1) and (8×2) phase of the Si(111)-In nanowire array. The stable phase is indicated. The inset shows enlarged the entropy difference calculated by neglecting the electronic contributions and by restricting the BZ sampling to the Γ-point.

The calculations are performed at the equilibrium lattice constant. From calculations where the measured lattice expansion has been taken into account, the corresponding error is estimated to be of the order of 0.1 meV per surface In atom. The sampling of the phonon dispersion curves is another crucial point. It is performed here by using only the Γ and the X point of the (8×2) BZ. However, as shown in the inset of Fig. 1.2, further restricting of the sampling to the Γ point results in an energy shift of less than 0.3 meV, indicating that the unit cell is large enough to compensate for poor BZ sampling. In chapter 3 (cf. Publ. [14]) it was demonstrated that the energetics of the In nanowires depends sensitively on the functional used to model the electron exchange and correlation energy and the treatment of the In $4d$ electrons. The inclusion of the In $4d$ states and/or the usage of the generalized gradient rather than the local density approximation is found to result in typical (maximum) frequency shifts of $\pm 2(4)$ cm^{-1}. This affects the vibrational free energy by at most 1 meV per surface In atom at 130 K.

Fig. 1.2 presents the free energy difference ΔF between the Si(111)-(4×1)-In and Si(111)-(8×2)-In phases. ΔF vanishes at 128.5 K if only the vibrational entropy is

8.1. Entropy explains the phase transition

taken into account. Additional consideration of the electronic entropy lowers the calculated phase transition temperature to 125 K. At this temperature, the vibrational and electronic entropy is large enough to compensate for the lower total energy of the insulating (8×2) phase compared to the metallic (4×1) phase. The calculated phase transition temperature is slightly above the experimental value of about 120 K. However, given the approximations and uncertainties discussed above, the agreement between theory and experiment should be considered to be fortuitously close.

The present calculations show that the phase transition is caused by the gain in (mainly vibrational) entropy that for higher temperatures overcompensates the gain in band-structure energy realized by transforming the 1D-metallic In zigzag chains into the semiconducting In hexagons. Is it possible to trace the change in vibrational entropy to the frequency shift of a few illustrative modes? Due to the reduced symmetry of the hexagon structure, the phase transition results in modified phonon eigenvectors. This complicates the one-to-one comparison of the phonon modes. However, a general trend

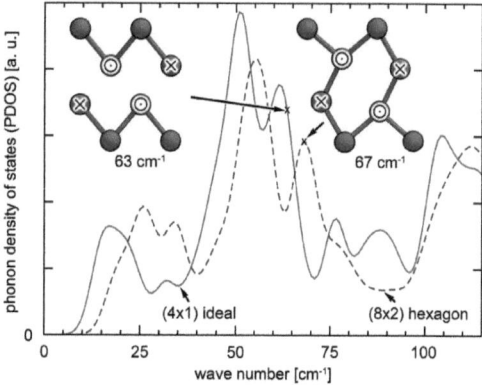

Figure 1.3: *Phonon density of states calculated for the (4×1) and (8×2) phase of the Si(111)-In nanowire array (4 cm^{-1} broadening). The inset shows a specific displacement pattern that hardly changes upon the phase transition but shifts in frequency. Arrows (heads/tails) indicate up/down movements.*

to higher surface phonon frequencies upon hexagon formation is clearly observed. This can be seen from most values in Tab. 1.1 – with the shear mode as a notable exception – as well as from the comparison of the respective phonon density of states shown in Fig. 1.3. The present calculations essentially confirm earlier experimental work that states "all major modes of the (4×1) surface are found in the (8×2) spectra, though blueshifted" [46]. A typical example is shown as inset in Fig. 1.3. The eigenvector corresponding to the alternating up and down movements of the In atoms hardly changes upon the (4×1) – (8×2) phase transition. The according frequency, however, goes up from 63 to 67 cm^{-1}. This shift in frequency is easily understood from the formation of additional In-In bonds upon hexagon formation, resulting in larger force constants.

In summary, free energy calculations based on density functional theory are performed that explain the (4×1) – (8×2) phase transition of the Si(111)-In nanowire array in terms of a subtle interplay between the lower total energy of the insulating In hexagon structure and the larger vibrational and electronic entropy of the less tightly bound and metallic In zigzag chain structure at finite temperatures. Both the (4×1) and (8×2) phases are stable and well-defined structural phases, characterized, e.g., by phase-specific phonon modes. Soft shear and rotary vibrations are identified that transform between the In zigzag chains stable at room temperature and the hexagons formed at low temperatures. The present work resolves the discrepancies between many experiments and the dynamic fluctuation model, where the (4×1) reconstruction is interpreted as time-averaged superposition of (8×2) structures and clarifies the long-standing issue of the temperature-induced metal-insulator transition in one of the most intensively investigated quasi-1D electronic systems. The mechanism revealed here is expected to apply to many more quasi-1D systems with intriguing phase transitions, e.g., Au nanowires on high-index silicon surfaces.

Bibliography

[1] G. Moore, *Cramming more components onto integrated circuits*, Electronics Magazine, pp. 4, 19th April (1965)

[2] R. Kurzweil, *The law of accelerating returns*, Kurzweil Applied Intelligence, http://www.kurzweilai.net

[3] M. Elbing, R. Ochs, M. Koentopp, M. Fischer, C. v. Hänisch, F. Weigend, F. Evers, H. B. Weber, M. Mayor, *A single-molecule diode*, Proc. Nat. Acad. Sci. USA **102**, 8815 (2005)

[4] C. Joachim, J. K. Gimzewski, *An electromechanical amplifier using a single molecule*, Chem. Phys. Lett. **265**, 353 (1997)

[5] A. K. Geim, K. S. Novoselov, *The rise of graphene*, Nature Materials **6**, 183 (2007)

[6] C. J. Muller, J. M. van Ruitenbeek, L. J. de Jongh, *Conductance and supercurrent discontinuities in atomic-scale metallic constrictions of variable width*, Phys. Rev. Lett. **69**, 140 (1992)

[7] D. Sanchez-Portal, E. Artacho, J. Junquera, P. Ordejon, A. Garcia, J. M. Soler, *Stiff Monatomic Gold Wires with a Spinning Zigzag Geometry*, Phys. Rev. Lett. **83**, 3884 (1999)

[8] N. Agrait, A. L. Yeyati, J. M. van Ruitenbeek, *Quantum properties of atomic-sized conductors*, Phys. Rep. **377**, 81 (2003)

[9] D. Wang, B. Sheriff, J. R. Heath, *Silicon p-FETs from ultrahigh density nanowire arrays*, Nano Lett. **6**, 1096 (2006)

[10] L. Chua, *Memristor – the missing circuit element*, IEEE Transactions on Circuit Theory **18**, 5 (1971)

[11] D. Strukov, G. Snider, D. Stewart, R. Williams, *The missing memristor found*, Nature **453**, 80 (2008)

[12] J. Yang, M. Pickett, X. Li, D. Ohlberg, D. Stewart, R. Williams, *Memristive switching mechanism for metal/oxide/metal nanodevices*, Nature Nanotechnology **3**, 429 (2008)

[13] R. Landauer, *Irreversibility and heat generation in the computing process*, IBM J. of Research & Development **17**, 525 (1973)

[14] R. Feynman, *Quantum mechanical computers*, Foundations of Physics **16**, 6 (1986)

[15] C. H. Bennett, *The thermodynamics of computation – a review*, Int. J. of Theor. Phys. **21**, 905 (1982)

[16] M. P. Frank, T. F. Knight, *Ultimate theoretical models of nanocomputers*, Nanotechnology **9**(3), 162 (1998)

[17] M. P. Frank, *Physical limits of computing*, Comp. in Sci. & Eng. **4**(3), 16 (2002)

[18] S. Younis, T. Knight, *Asymptotically zero energy computing using split-level charge recovery logic*, PhD thesis, Massachusetts Institute of Technology 1994

[19] J. M. Luttinger, *An exactly soluble model of a many-Fermion system*, J. Math. Phys. **4**, 1154 (1963)

[20] R. Peierls, *More surprises in theoretical physics*, Princeton University Press (1991)

[21] W. Tremel, E. W. Finck, *Ladungsdichtewellen: Elektrische Leitfähigkeit*, Chemie in unserer Zeit **38**, 326 (2004)

[22] J. J. Lander, J. Morrison, *Surface Reactions of Silicon (111) with Aluminum and Indium*, J. Appl. Phys. **36**, 1706 (1965)

[23] T. Abukawa, M. Sasaki, F. Hisamatsu, T. Goto, T. Kinoshita, A. Kakizaki, S. Kono, *Surface electronic structure of a single-domain Si(111)4×1-In surface: a synchrotron radiation photoemission study*, Surf. Sci. **325**, 33 (1995)

[24] I. G. Hill, A. B. McLean, *Metallicity of In chains on Si(111)*, Phys. Rev. B **56**, 15725 (1997)

[25] O. Bunk, G. Falkenberg, J. H. Zeysing, L. Lottermoser, R. L. Johnson, M. Nielsen, F. Berg-Rasmussen, J. Baker, R. Feidenhans'l, *Structure determination of the indium-induced Si(111)-(4×1) reconstruction by surface x-ray diffraction*, Phys. Rev. B **59**, 12228 (1999)

[26] F. Owman, P. Mårtensson, *STM study of hydrogen exposure of the Si(111)$\sqrt{(3)} \times \sqrt{3}$-In surface*, Surf. Sci **359**, 122 (1996)

[27] T. Abukawa, M. Sasaki, F. Hisamatsu, N. Nakamura, T. Kinoshita, A. Kakizaki, T. Goto, S. Kono, *Core-level photoemission study of the Si(111)4 × 1-In surface*, J. Electron. Spectrosc. Relat. Phenom. **80**, 233 (1996)

[28] N. Nakamura, K. Anno, S. Kono, *Structure analysis of the single-domain Si(111)4 × 1-In surface by µ-probe Auger electron diffraction and µ-probe reflection high energy electron diffraction*, Surf. Sci. **256**, 129 (1991)

[29] J.-H. Cho, D.-H. Oh, K. S. Kim, L. Kleinman, *Weakly correlated one-dimensional indium chains on Si(111)*, Phys. Rev. B **64**, 235302 (2001)

[30] H. W. Yeom, S. Takeda, E. Rotenberg, I. Matsuda, K. Horikoshi, J. Schaefer, C. M. Lee, S. D. Kevan, T. Ohta, T. Nagao, S. Hasegawa, *Instability and Charge Density Wave of Metallic Quantum Chains on a Silicon Surface*, Phys. Rev. Lett. **82**, 4898 (1999)

[31] S. J. Park, H. W. Yeom, J. R. Ahn, I.-W. Lyo, *Atomic-Scale Phase Coexistence and Fluctuation at the Quasi-One-Dimensional Metal-Insulator Transition*, Phys. Rev. Lett. **95**, 12601 (2005)

[32] C. Kumpf, O. Bunk, J. H. Zeysing, Y. Su, M. Nielsen, R. L. Johnson, R. Feidenhans'l, K. Bechgaard, *Low-temperature structure of indium quantum chains on silicon*, Phys. Rev. Lett. **85**, 4916 (2000)

[33] T. Uchihashi, U. Ramsperger, *Electron conduction through quasi-one-dimensional indium wires on silicon*, Appl. Phys. Lett. **80**, 22 (2002)

[34] C. González, J. Ortega, F. Flores, *Metal-insulator transition in one-dimensional In-chains on Si(111): combination of a soft shear distortion and a double-band Peierls instability*, New J. Phys. **7**, 100 (2005)

[35] C. González, F. Flores, J. Ortega, *Soft phonon, dynamical fluctuations and reversible phase transition: In chains on Silicon*, Phys. Rev. Lett. **96**, 136101 (2006)

[36] J. R. Ahn, J. H. Byun, H. Koh, E. Rotenberg, S. D. Kevan, H. W. Yeom, *Mechanism of Gap Opening in a Triple-Band Peierls System: In Atomic Wires on Si*, Phys. Rev. Lett **93**, 106401 (2004)

[37] C. González, J. Guo, J. Ortega, F. Flores, H. H. Weitering, *Mechanism of the Band Gap Opening across the Order-Disorder Transition of Si(111)(4×1)-In*, Phys. Rev. Lett. **102**, 115501 (2009)

[38] J.-H. Cho, J.-Y. Lee, *First-principles calculation of the atomic structure of one-dimensional indium chains on Si(111): Convergence to a metastable structure*, Phys. Rev. B **76**, 033405 (2007)

[39] H. W. Yeom, *Comment on "Soft Phonon, Dynamical Fluctuations, and a Reversible Phase Transition: Indium Chains on Silicon"*, Phys. Rev. Lett. **97**, 189701 (2006)

[40] M. Hashimoto, Y. Fukaya, A. Kawasuso, A. Ichimiya, *Quasi-one-dimensional In atomic chains on Si(111) at low temperature studied by reflection high-energy positron diffraction and scanning tunneling microscopy*, Appl. Surf. Sci. **254**, 7733 (2008)

[41] Y. Fukaya, M. Hashimoto, A. Kawasuso, A. Ichimiya, *Surface structure of Si(111)-(8×2)-In determined by reflection high-energy positron diffraction*, Surf. Sci. **602**, 2448 (2008)

[42] S. J. Park, H. W. Yeom, S. H. Min, D. H. Park, I.-W. Lyo, *Direct Evidence of the Charge Ordered Phase Transition of Indium Nanowires on Si(111)*, Phys. Rev. Lett. **93**, 106402 (2004)

[43] T. Kanagawa, R. Hobara, I. Matsuda, T. Tanikawa, A. Natori, S. Hasegawa, *Anisotropy in conductance of a quasi-one-dimensional metallic surface state measured by a square micro-four-point probe method*, Phys. Rev. Lett. **91**, 036805 (2003)

[44] T. Tanikawa, I. Matsuda, T. Kanagawa, S. Hasegawa, *Surface state electrical conductivity at a metal-insulator transition on silicon*, Phys. Rev. Lett. **93**, 016801 (2004)

[45] H. W. Yeom, K. Horikoshi, H. M. Zhang, K. Ono, R. I. G. Uhrberg, *Nature of the broken symmetry phase of the one-dimensional metallic In/Si(111) surface*, Phys. Rev. B **65**, 241307 (2002)

[46] K. Fleischer, S. Chandola, N. Esser, W. Richter, J. F. McGilp, *Surface phonons of the Si(111):In-(4×1) and (8×2) phases*, Phys. Rev. B **76**, 205406 (2007)

[47] F. Bechstedt, A. Krivosheeva, J. Furthmüller, A. A. Stekolnikov, *Vibrational properties of the quasi-one-dimensional In/Si(111)-(4×1) system*, Phys. Rev. B **68**, 193406 (2003)

[48] Y. J. Sun, S. Agario, S. Souma, K. Sugawara, Y. Tago, T. Sato, T. Takahashi, *Cooperative structural and Peierls transition of indium chains on Si(111)*, Phys. Rev. B **77**, 125115 (2008)

[49] F. Pedreschi, J. D. O'Mahony, P. Weightman, J. R. Power, *Evidence of electron confinement in the single-domain (4×1)-In superstructure on vicinal Si(111)*, Appl. Phys. Lett. **73**, 2152 (1998)

[50] J. F. McGilp, *Optical response of low-dimensional In nanostructures grown by self-assembly on Si surfaces*, phys. stat. sol. A **188**, 1361 (2001)

[51] K. Fleischer, S. Chandola, N. Esser, W. Richter, J. F. McGilp, *Phonon and polarized reflectance spectra from Si(111)-(4×1)In: Evidence for a charge-density-wave driven phase transition*, Phys. Rev. B **67**, 235318 (2003)

[52] S. Wang, W. Lu, W. G. Schmidt, J. Bernholc, *Nanowire-induced optical anisotropy of the Si(111)-In surface*, Phys. Rev. B **68**, 035329 (2003)

[53] S. V. Ryjkov, T. Nagao, V. G. Lifshits, S. Hasegawa, *Phase transition and stability of Si(111)-8×'2'-In surface phase at low temperatures*, Surf. Sci. **488**, 15 (2001)

[54] S. S. Lee, J. R. Ahn, N. D. Kim, J. H. Min, C. G. Hwang, J. W. Chung, H. W. Yeom, S. V. Ryjkov, S. Hasegawa, *Adsorbate-induced pinning of a charge-density wave in a quasi-1D metallic chain: Na on the In/Si(111)-(4×1) surface*, Phys. Rev. Lett **88**, 196401 (2002)

[55] J.-H. Cho, D.-H. Oh, L. Kleinman, *Theoretical study of Na adsorption on top of In chains on the Si(111) surface*, Phys. Rev. B **66**, 075423 (2002)

[56] H. Shim, S.-Y. Yu, W. Lee, J.-Y. Koo, G. Lee, *Control of phase transition in quasi-one-dimensional atomic wires by electron doping*, Appl. Phys. Lett. **94**, 231901 (2009)

[57] Y. Terada, S. Yoshida, A. Okubo, K. Kanazawa, M. Xu, O. Takeuchi, H. Shigekawa, *Optical doping: Active control of metal-insulator transition in nanowire*, Nano Lett. **8**, 3577 (2008)

[58] G. Lee, S.-Y. Yu, H. Kim, J.-Y. Koo, *Defect-induced perturbation on Si(111)4×1-In: Period-doubling modulation and its origin*, Phys. Rev. B **70**, 121304 (2004)

[59] S-Y. Yu, D. Lee, H. Kim, J.-Y. Koo, G. Lee, *Adsorption of H atoms on a Si(111)4×1-In surface*, J. Korean Phys. Soc. **48**, 1338 (2006)

[60] G. Lee, S-Y. Yu, D. Lee, H. Kim, J.-Y. Koo, *A new period-doubled modulation on the In/Si(111)4×1 surface induced by defects*, Jap. J. Appl. Phys. **45**, 2087 (2006)

[61] M. Hupalo, T.-L. Chan, C. Z. Wang, K.-M. Ho, M. C. Tringides, *Interplay between indirect interaction and charge-density wave in Pb-adsorbed In(4×1)-Si(111)*, Phys. Rev. B **76**, 045415 (2007)

[62] S. S. Lee, S. Y. Shin, C. G. Hwang, J. W. Chung, *Observation of impurity-driven metal-insulator transitions in quasi-1D nanowires on a (Na, Li)-added In/Si(111)-4×1 surface*, J. Korean, Phys. Soc. **53**, 3667 (2008)

[63] C. Liu, T. Uchihashi, T. Nakayama, *Self-alignment of Co adatoms on In atomic wires by quasi-one-dimensional electron-gas-mediated interactions*, Phys. Rev. Lett **101**, 146104 (2008)

[64] G. Lee, S.-Y. Yu, H. Shim, W. Lee, J.-Y. Koo, *Roles of defects induced by hydrogen and oxygen on the structural phase transition of Si(111)4×1-In*, Phys. Rev. B **80**, 075411 (2009)

[65] M. Xu, A. Okada, S. Yoshida, H. Shigekawa, *Self-organization of In nanostructures on Si surfaces*, Appl. Phys. Lett. **94**, 073109 (2009)

[66] M. Born, J. R. Oppenheimer, *Zur Quantentheorie der Molekeln*, Annalen der Physik **389**, 457-484 (1927)

[67] A. Szabo, N. S. Ostlund, *Modern quantum chemistry: Introduction to advanced electronic structure theory*, McGraw-Hill (1989)

[68] G. P. Srivastava, D. Weaire, *The theory of the cohesive energy of solids*, Advances in Physics **36**, 463-517 (1987)

[69] K. Capelle, *A Bird's eye view of density functional theory*, Proceedings of the VIIIth Brazilian summer school on electronic structure **10**, S. 1 (2003)

[70] R. G. Parr, W. Yang, *Density-functional theory of atoms and molecules*, Oxford University Press (1989)

[71] P. Hohenberg, W. Kohn, *Inhomogeneous electron gas*, Phys. Rev. B **136**, B864 (1964)

[72] W. Kohn, L. J. Sham, *Self-consistent equations including exchange and correlation effects*, Phys. Rev. **140**, A1133 (1965)

[73] J. Harris, *Adiabatic-connection approach to Kohn-Sham theory*, Phys. Rev. A **29**, 1648 (1984)

[74] U. v. Barth, L. Hedin, *Local exchange-correlation potential for spin polarized case .1.*, J. Phys. C **5**, 1629 (1972)

[75] O. Gunnarson, B. Lundqvist, *Exchange and correlation in atoms, molecules and solids by the spin-density-functional formalism*, Phys. Rev. B **13**, 4274 (1976)

[76] D. M. Ceperly, B. J. Alder, *Ground state of the electron gas by a stochastic method*, Phys. Rev. Lett. **45**, 566 (1980)

[77] E. Wigner, *On the interactions of electrons in metals*, Phys. Rev. **46**, 1002 (1932)

[78] M. Gell-Mann, K. A. Brueckner, *Correlation of an electron gas at high density*, Phys. Rev. **106**, 364 (1957)

[79] J. P. Perdew, A. Zunger, *Self-interaction correction to density functional approximations for many-electron systems*, Phys. Rev. B **23**, 50 48 (1981)

[80] S. H. Vosko, L. Wilk, M. Nusair, *Accurate spin-dependent electron liquid correlation energies for local spin-density calculations – A critical analysis*, Can. J. Phys. **48**, 1200 (1980)

[81] Y. Wang, J. P. Perdew, *Correlation-hole of the spin-polarized electron gas, with exact small-wave vector and high density scaling*, Phys. Rev. B **44**, 13298 (1991)

[82] J. P. Perdew, J. A. Chevari, S. H. Vosko, K. A. Jackson, M. R. Pederson, D. J. Singh, C. Fiolhais, *Atoms, molecules, solids and surfaces: Applications of the generalized gradient approximation for exchange and correlation*, Phys. Rev B **46**, 6671 (1992)

[83] J. P. Perdew, K. Burke, M. Ernzerhof, *Generalized gradient expansion made simple*, Phys. Rev. Lett. **77**, 3865 (1996)

[84] F. Ortmann, F. Bechstedt, W. G. Schmidt, *Semiempirical van der Waals correction to the density functional description of solids and molecular structures*, Phys. Rev. B **73**, 205101 (2006)

[85] J. P. Perdew, L. G. Parr, M. Levy, J. L. Balduz, *Density-functional theory for fractional particle number: Derivative discontinuities of the energy*, Phys. Rev. Lett. **49**, 1691 (1982)

[86] A. Baldereschi, *Mean-value point in the Brillouin zone*, Phys. Rev. B **7**, 5212 (1973)

[87] V. K. Bashenov, M. Bardashova, A. M. Mutal, *Baldereschi points for some noncubic lattices*, phys. stat. sol. (b) **80**, 89 (1977)

[88] D. J. Chadi, M. L. Cohen, *Special points in the Brillouin zone*, Phys. Rev. B **8**, 5747 (1973)

[89] P. J. Lin-Chung, *Many special point scheme for noncubic lattices*, phys. stat. sol. (b) **85**, 743 (1978)

[90] H. J. Monkhorst, J. D. Pack, *Special points for Brillouin zone integration*, Phys. Rev. B **13**, 5188 (1976)

[91] M. Methfessel, A. Paxton, *High-precision sampling for Brillouin-zone integration in metals*, Phys. Rev. B **40**, 3616 (1989)

[92] N. Marzari, D. Vanderbilt, A. de Vita, M. C. Payne, *Thermal contraction and disordering of the Al(110) surface*, Phys. Rev. Lett. **82**, 3296 (1999)

[93] J. C. Philips, L. Kleinman, *New method for calculating wave functions in crystals and molecules*, Phys. Rev. B **116**, 287 (1959)

[94] D. R. Hamann, M. Schlüter, C. Chiang, *Norm-conserving pseudopotentials*, Phys. Rev. Lett. **43**, 1494 (1979)

[95] G. B. Bachelet, D. R. Hamann, M. Schlüter, *Pseudopotentials that work: From H to Pu*, Phys. Rev. B **26**, 4199 (1982)

[96] N. Troullier, J. L Martins, *Efficient pseudopotentials for plane-wave calculations*, Phys. Rev. B **43**, 1993 (1991)

[97] L. Kleinman, D. M. Bylander, *Efficacious form for model pseudopotentials*, Phys. Rev. Lett. **48**, 1425 (1982)

[98] D. Vanderbilt, *Soft self-consistent pseudopotentials in a generalized eigenvalue formalism*, Phys. Rev. B Rapid Comm. **41**, 7892 (1990)

[99] P. E. Blöchl, *Projector augmented wave method*, Phys. Rev. B **50**, 17953 (1994)

[100] G. Kresse, D. Joubert, *From ultrasoft pseudopotentials to the projector augmented wave method*, Phys. Rev. B **59**, 1758 (1999)

[101] R. Feynman, *Forces in molecules*, Phys. Rev. B **56**, 340 (1937)

[102] A. C. Hurley, *The electrostatic calculation of molecular energies. 1) Methods of calculating molecular energies*, Proc. R. Soc. London, Ser. A **226**, 170 (1954)

[103] M. D. Ventra, S. T. Pantelides, *Hellmann-Feynman theorem and the definition of forces in quantum time-dependent and transport problems*, Phys. Rev. B **61**, 16207 (2000)

[104] *Vienna ab initio simulation package (VASP)*:
http://www.cms.mpi.univie.ac.at/vasp

[105] E. N. Economou, *Green's functions in quantum physics*, Springer-Verlag, Berlin (1979)

[106] S. Datta, *Electronic transport in mesoscopic systems*, Cambridge University Press, Cambridge, UK (1995)

[107] S. Datta, *Quantum transport: Atom to transistor*, Cambridge University Press, Cambridge, UK (2005)

[108] R. Feynman, R. Leighton, M. Sands, *Lectures on physics, vol. I*, Addison-Wesley, Amsterdam (1998)

[109] R. Feynman, R. Leighton, M. Sands, *Lectures on physics, vol. II*, Addison-Wesley, Amsterdam (1998)

[110] R. Feynman, R. Leighton, M. Sands, *Lectures on physics, vol. III*, Addison-Wesley, Amsterdam (1998)

[111] T. J. Drummond, W. T. Masselink, H. Morkoc, *Modulation-doped GaAs/(Al, Ga)AS heterojunction field-effect transistors: MODFET's*, Proc. IEEE **74**, 773 (1986)

[112] M. R. Melloch, *Molecular beam epitaxy for high electron mobility modulation-doped two-dimensional electron gases*, Thin Solid Films **231**, 74 (1993)

[113] S. Washburn, R. A. Webb, *Aharonov-Bohm effect in normal metal: Quantum coherence and transport*, Adv. Phys. **35**, 375 (1986)

[114] B. Ötzel, *Diploma thesis: Ab-initio Untersuchungen zu Quantentransport auf molekularen Skalen*, Friedrich Schiller Universität Jena, Institut für Festkörpertheorie und -optik (2008)

[115] R. Landauer, *Spatial variation of currents and fields due to localized scatterers in metallic conduction*, IBM J. Res. Dev. **32**, 306 (1988)

[116] R. Landauer, *Conductance from transmission: common sense points*, Physica Scripta **T42**, 110 (1992)

[117] A. Szafer, A. D. Stone, *Theory of quantum conduction through a constriction*, Phys. Rev. Lett. **62**, 300 (1989)

[118] D. S. Fisher, P. A. Lee, *Relation between conductivity and transmission matrix*, Phys. Rev. B **23**, 6851 (1981)

[119] D. H. Lee, J. D. Joannopoulos *Simple scheme for surface-band calculations. I*, Phys. Rev. B **23**, 4988 (1981)

[120] D. H. Lee, J. D. Joannopoulos *Simple scheme for surface-band calculations. II. The Green's function*, Phys. Rev. B **23**, 4997 (1981)

[121] M. Lopez-Sancho, J. Lopez-Sancho, J. Rubio *Quick iterative scheme for the calculation of transfer matrices: application to Mo (100)*, J. Phys. F: Metal Physics **14**, 1205 (1984)

[122] M. Lopez-Sancho, J. Lopez-Sancho, J. Sancho, J. Rubio *Highly convergent schemes for the calculation of bulk and surface Green functions*, J. Phys. F: Metal Physics **15**, 851 (1985)

[123] F. Garcia-Moliner, V. Velasco, *Theory of Single and Multiple Interfaces*, World Scientific, Singapore (1992)

[124] F. Garcia-Moliner, V. Velasco, *Matching methods for single and multiple interfaces: Discrete and continuous media*, Phys. Rep. **200**, 83 (1991)

[125] M. B. Nardelli, *Electronic transport in extended systems: Application to carbon nanotubes*, Phys. Rev. B **60**, 7828 (1999)

[126] N. Marzari, D. Vanderbilt, *Maximally localized generalized Wannier functions for composite energy bands*, Phys.Rev. B **56**, 12847 (1997)

[127] I. Souza, N. Marzari, D. Vanderbilt, *Maximally localized Wannier functions for entangled energy bands*, Phys. Rev. B **65**, 035109 (2001)

[128] A. Calzolari, N. Marzari, I. Souza, M. B. Nardelli, *Ab initio transport properties of nanostructures from maximally localized Wannier functions*, Phys. Rev. B **69**, 035108 (2004)

[129] J. M. Foster, S. F. Boys, *Quantum Variational Calculations for a Range of CH_2 Configurations*, Rev. Mod. Phys. **32**, 305 (1962)

[130] E. Blount, *Formalisms of band theory*, Sol. Stat. Phys. **13**, 305 (1962)

[131] *Plane-Wave Self-consistent field (PWScf)* program package:
http://www.quantum-espresso.org

[132] *Wannier Transport (WanT)* program package:
http://www.wannier-transport.org

[133] D. E. Aspnes, *Real-time optical analysis and control of semiconductor epitaxy: progress and opportunity*, Solid State Commun. **101**, 85 (1997)

[134] P. Weightman, D. S. Martin, R. J. Cole, T. Farrell, *Reflection anisotropy spectroscopy*, Rep. Prog. Phys. **68**, 1251 (2005)

[135] W. G. Schmidt, *Calculation of reflectance anisotropy for semiconductor surface exploration*, phys. stat. sol. (b) **242**, 2751 (2005)

[136] W. G. Schmidt, F. Bechstedt, J. Bernholc, *Understanding reflectance anisotropy: Surface-state signatures and bulk-related features*, J. Vac. Sci. Technol. B **18**, 2215 (2000)

[137] W. G. Schmidt, K. Seino, P. H. Hahn, F. Bechstedt, W. Lu, S. Wang, J. Bernholc, *Calculation of surface optical properties: from qualitative understanding to quantitative predictions*, Thin Solid Films **455/456**, 764 (2004)

[138] W. G. Schmidt, F. Fuchs, A. Hermann, K. Seino, F. Bechstedt, R. Paßmann, M. Wahl, M. Gensch, K. Hinrichs, N. Esser, S. Wang, W. Lu, J. Bernholc, *Oxidation- and organic-molecule-induced changes of the Si surface optical anisotropy: ab initio predictions*, J. Phys.: Condens. Matter **16**, S4323 (2004)

[139] J. D. E. McIntyre, D. E. Aspnes, *Differential reflection spectroscopy of very thin surface films*, Surf. Sci. **24**, 417 (1971)

[140] A. Bagchi, R. G. Barrera, A. K. Rajagopal, *Perturbative approach to the calculation of the electric field near a metal surface*, Phys. Rev. B **20**, 4824 (1979)

[141] R. Del Sole, *Microscopic theory of optical properties of crystal surfaces*, Solid State Commun. **37**, 537 (1981)

[142] R. Del Sole, E. Fiorino, *Macroscopic dielectric tensor at crystal surfaces*, Phys. Rev. B **29**, 4631 (1984)

[143] F. Manghi, R. Del Sole, A. Selloni, E. Molinari, *Anisotropy of surface optical properties from first-principles calculations*, Phys. Rev. B **41**, 9935 (1990)

[144] B. Adolph, V. I. Gavrilenko, K. Tenelsen, F. Bechstedt, R. Del Sole, *Nonlocality and many-body effects in the optical properties of semiconductors*, Phys. Rev. B **53**, 9797 (1996)

[145] H. Ehrenreich, M. H. Cohen, *Self-Consistent Field Approach to the Many-Electron Problem*, Phys. Rev. **115**, 786 (1959)

[146] S. L. Adler, *Quantum Theory of the Dielectric Constant in Real Solids*, Phys. Rev. **126**, 413 (1962)

[147] N. Wiser, *Dielectric Constant with Local Field Effects Included*, Phys. Rev. **129**, 62 (1963)

[148] G. Grosso, G. P. Parravicini, *Solid State Physics*, Academic Press, San Diego, California (2003)

[149] N. W. Ashcroft, N. W. Mermin, *Solid State Physics*,

[150] F. Jensen, *Introduction to computational chemistry*, John Wiley & Sons Ltd., Chichester, UK (1999)

[151] A. Marini, G. Onida. R. Del Sole, *Plane-wave DFT-LDA calculation of the electronic structure and absorption spectrum of copper*, Phys. Rev. B **64**, 195125 (2001)

[152] P. Monachesi, M. Palumno, R. Del Sole, *Reflectance anisotropy spectra of Cu and Ag (110) surfaces from* ab initio *theory*, Phys. Rev. B **64**, 115421 (2001)

[153] L. Hedin, *New Method for calculating the one-particle Green's function with application to the electron gas problem*, Phys. Rev. **139**, A796 (1965)

[154] L. Hedin, S. Lundquist, *Solid State physics* **23**, Academic Press, New York (1969)

[155] F. Aryasetiawan, O. Gunnarson, *The GW method*, Rep. Prog. Phys. **61**, 237 (1998)

[156] J. Schwinger, *On the Green's Functions of Quantized Fields*, Proc. Natl. Acad. Sci. **37**, 452 (1951)

[157] P. C. Martin, J. Schwinger, *Theory of Many-Particle Systems. I*, Phys. Rev **115**, 1342 (1959)

[158] G. Onida, L. Reining, A. Rubio, *Electronic excitations: density-functional versus many-body Green's function approaches*, Rev. Mod. Phys. **74**, 601 (2002)

[159] M. S. Hybertsen, S. G. Louie, *First-principles theory of quasi-particles: calculation of band gaps in semiconductors and insulators*, Phys. Rev. Lett. **55**, 1418 (1985)

[160] M. S. Hybertsen, S. G. Louie, *Electron correlation in semiconductors and insulators: band gaps and quasi-particle energies*, Phys. Rev. B **34**, 5390 (1986)

[161] M. S. Hybertsen, S. G. Louie, *Model dielectric matrices for quasiparticle self-energy calculations*, Phys. Rev. B **37**, 2733 (1988)

[162] F. Bechstedt, R. Del Sole, G. Cappellini, L. Reining, *An efficient method for calculating quasi-particle energies in semiconductors*, Solid State Commun. **84**, 765 (1992)

[163] W. G.Schmidt, S. Glutsch, P. H. Hahn, F. Bechstedt, *Efficient $\mathcal{O}(N^2)$ method to solve the Bethe-Salpeter equation*, Phys. Rev. B **67**, 085307 (2003)

[164] L. J. Sham, T. M. Rice, *Many-Particle Derivation of the Effective-Mass Equation for the Wannier Exciton*, Phys. Rev **144**, 708 (1966)

[165] W. Hanke, L. J. Sham, *Many-particle effects in the optical spectrum of a semiconductor*, Phys. Rev B **21**, 4656 (1980)

[166] H. Stolz, *Einführung in die Vielelektronentheorie der Kristalle*, Akademie-Verlag Berlin (1974)

[167] S. Albrecht, L. Reining, R. Del Sole, G. Onida, *Ab Initio Calculation of Excitonic Effects in the Optical Spectra of Semiconductors*, Phys. Rev. Lett. **80**, 4510 (1998)

[168] M. Rohlfing, S. G. Louie, *Excitonic Effects and the Optical Absorption Spectrum of Hydrogenated Si Clusters*, Phys. Rev. Lett. **80**, 3320 (1998)

[169] A. Dhar, A. Mansingh, *Optical properties of reduced lithium niobate single crystals*, J. Appl. Phys. **68**, 5804 (1990)

[170] R. F. Schaufele, M. J. Weber, *Raman Scattering by Lithium Niobate*, Phys. Rev. **152**, 705 (1966)

[171] Z. Jiangou, Z. hipin, X. Dingquan, W. Xiu, X. Guanfeng, *Optical absorption properties of doped lithium niobate crystals*, J. Phys.: Condens. Matter **4**, 2977 (1992)

[172] D. Redfield, W. J. Burke, *Optical absorption edge of $LiNbO_3$*, J. Appl. Phys. **45**, 4566 (1974)

[173] S. Kase, K. Ohi, *Optical absorption and interband faraday rotation in LiTaO$_3$ and LiNbO$_3$*, Ferroelectrics **8**, 419, (1974)

[174] M. Veithen, P. Ghosez, *First-principles study of the dielectric and dynamical properties of lithium niobate*, Phys. Rev. B **65**, 214302 (2002)

[175] I. V. Kityk, M. Makowska-Janusik, M. D. Fontana. M. Aillerie, F. Abdi, *Band structure treatment of the influence of nonstoichiometric defects on optical properties in LiNbO$_3$*, J. Appl. Phys. **90**, 5542 (2001)

[176] W. Y. Ching, Z. Q. Gu, Y. N. Xu, *First-principles calculation of the electronic and optical properties of LiNbO$_3$*, Phys Rev B **50**, 1992 (1994)

[177] F. Bechstedt, K. Seino, P. H. Hahn, W. G. Schmidt, *Quasiparticle bands and optical spectra of highly ionic crystals: AlN and NaCl*, Phys. Rev. B **72**, 245114 (2005)

[178] A. M. Mamedov, M. A. Osman, L. C. Hajieva, *VUV reflectivity of LiNbO$_3$ and LiTaO$_3$ single crystals*, Appl. Phys. A: Solids Surf. **34**, 189 (1984)

[179] E. Wiesendanger, G. Güntherodt, *Optical anisotropy of LiNbO$_3$ and KNbO$_3$ in the interband transition region*, Solid State Commun. **14**, 303 (1974)

[180] R. Kosloff, *Time-dependent quantum-mechanical methods for molecular dynamics*, J. Phys. Chem. **92**, 2087 (1988)

[181] E. Runge, E. Gross, *Density-functional theory for time-dependent systems*, Phys. Rev. Lett. **52**, 997 (1984)

[182] *Dielectric properties (DP) program package*, http://dp-code.org

[183] U. Häussermann, S. I. Simak, R. Ahuja, B. Johansson, S. Lidin, *The origin of the distorted close-packed elemental structure of Indium*, Angew. Chem. **38**, 2017 (1999)

[184] A. S. Mikhaylushkin, U. Häussermann, B. Johansson, and S. I. Simak, *Fluctuating lattice constants of indium under high pressure*, Phys. Rev. Lett. **92**, 195501 (2004)

[185] S. I. Simak, U. Häussermann, R. Ahuja, S. Lidin, B. Johansson, *Gallium and indium under high pressure*, Phys. Rev. Lett. **85**, 142 (2000)

[186] X. Lopez-Lozano, A. A. Stekolnikov, J. Furthmüller, F. Bechstedt, *Band structure and electron gas of In chains on Si(111)*, Surf. Sci. **589**, 77 (2005)

[187] R. H. Miwa, G. P. Srivastava, *Atomic geometry, electronic structure and image state for the Si(111)-In(4×1) nanowire*, Surf. Sci. **473**, 123 (2001)

[188] J.-H. Cho, J.-Y. Lee, L. Kleinman, *Electronic structure of one-dimensional indium chains on Si(111)*, Phys.Rev. B **71**, 081310 (2005)

[189] X. Lopez-Lozano, A. Krivosheeva, A. A. Stekolnikov, L. Meza-Montes, C. Noguez, J. Furthmüller, F. Bechstedt, *Reconstruction of quasi-one-dimensional In/Si(111) systems: Charge- and spin-density waves versus bonding*, Phys. Rev. B **73**, 035430 (2006)

[190] A. Calzolari, C. Cavazzoni, M. B. Nardelli, *Electronic and transport properties of artificial gold chains*, Phys. Rev. Lett. **93**, 096404

[191] K. Fleischer, S. Chandola, N. Esser, W. Richter, J. F. McGilp, *Reflectance anisotropy spectroscopy of Si(111)-(4×1)-In*, phys. stat. sol a **188**, 1411 (2001)

[192] K. Fleischer, S. Chandola, N. Esser, W. Richter, J. F. McGilp, W. G. Schmidt, S. Wang, W. Lu, J. Bernholc, *Atomic indium nanowires on Si(111): the (4×1)-(8×2) phase transition studied with reflectance anisotropy spectroscopy*, Appl. Surf. Sci. **234**, 302 (2004)

[193] P. H. Hahn, W. G. Schmidt, F. Bechstedt, *Bulk excitonic effects in surface optical spectra*, Phys. Rev. Lett. **88**, 016402 (2001)

[194] S. Nosé, *A unified formulation of the constant temperature molecular dynamics methods*, J. Chem. Phys. **81**, 511 (1984)

[195] S. Nosé, *Constant temperature molecular dynamics methods*, Prog. Theor. Phys. Suppl. **103**, 1 (1991)

[196] D. M. Bylander, L. Kleinman, *Energy fluctutations induced by the Nosé thermostat*, Phys. Rev. B **46**, 13756 (1992)

[197] N. Koch, *Master's Thesis: Temperaturabhängigkeit des Elektronentransports am Beispiel atomarer Ketten*

[198] F. Zhou, T. Maxisch, G. Ceder, *Configurational electronic entropy and the phase diagram of mixed-valence oxides: the case of Li_xFePO_4*, Phys. Rev. Lett. **97**, 155704 (2006)

[199] C. Wolverton, A. Zunger, *First-principles theory of short-range order, electronic excitations, and spin polarization in Ni-V and Pd-V alloys*, Phys. Rev. B **52**, 8813 (1995)

[200] D. M. C. Nicholson, G. M. Stocks, Y. Wang, W. A. Shelton, *Stationary nature of the density-functional free energy: application to accelerated multiple-scattering calculations*, Phys. Rev. B **50**, 14686 (1994)

[201] N. Shao, E. Wickstrom, B. Panchapakesan, *Nanotube-antibody biosensor arrays for the detection of circulating breast cancer cells*, Nanotechnology **19**, 465101 (2008)

[202] J. Kim, B. Lee, S. Hong, S. Sim, *Ultrasensitive carbon nanotube-based biosensors using antibody-binding fragments*, Analytical Biochemistry **381**, 193 (2008)

[203] M Valtiner, M Todorova, G Grundmeier, and J Neugebauer, Phys. Rev. Lett. **103**, 065502 (2009).

[204] S Riikonen and A Ayuela and D Sanchez-Portal, Surf. Sci. **600**, 3821 (2006).

[205] *X-window CRYstalline Structures and DENsities* (XCrysDen) program package, http://www.xcrysden.org

[206] *IBM Visualization Data Explorer* (OpenDX) program package, http://www.opendx.org

[207] *Visual Molecular Dynamics* (VMD) program package, http://www.ks.uiuc.edu/Research/vmd/

VMD material properties employed throughout the present work:
Edgy: A=0.1, D=0.7, Sp=1.0, Sh=0.1, O=1.0, Ol=4.0, Ow=0.75
Trans: A=0.1, D=0.9, Sp=1.0, Sh=0.2, O=0.66, Ol=4.0, Ow=0.5

Cover image settings:
AOA=1.0, AOD=0.4
Opaque: A=1.0, D=0.6, Sp=0.8, Sh=0.5, O=1.0, Ol=0.0, Ow=0.0
Diffuse: A=0.0, D=0.7, Sp=0.0, Sh=0.5, O=1.0, Ol=0.0, Ow=0.0
Trans: A=0.0, D=0.9, Sp=1.0, Sh=0.2, O=0.66, Ol=1.9, Ow=0.0

Publications

1. S. Wippermann, W. G. Schmidt, «*Entropy explains metal-insulator transition of Si(111)-In nanowire array*», Phys. Rev. Lett. **105**, 126102 (2010)

2. E. Speiser, S. Chandola, K. Hinrichs, M. Gensch, C. Cobet, S. Wippermann, W. G. Schmidt, F. Bechstedt, W. Richter, K. Fleischer, J. F. McGilp, N. Esser, «*Metal-insulator transition in Si(111)-(4x1)/(8x2)-In studied by optical spectroscopy*», Phys. Stat. Sol. B **247**, 2033 (2010)

3. S. Wippermann, W. G. Schmidt, F. Bechstedt, S. Chandola, K. Hinrichs, M. Gensch, N. Esser, K. Fleischer, J. F. McGilp, «*Optical anisotropy of Si(111)-(4×1)/(8×2)-In nanowires calculated from first-principles*», Phys. Stat. Sol. C **7**, 133 (2010)

4. S. Wippermann, W. G. Schmidt, P. Thissen, G. Grundmeier, «*Dissociative and molecular adsorption of water on $\alpha - Al_2O_3(0001)$*», Phys. Stat. Sol. C **7**, 137 (2010)

5. P. Thissen, G. Grundmeier, S. Wippermann, W. G. Schmidt, «*Water adsorption on the $\alpha - Al_2O_3(0001)$ surface*», Phys. Rev. B **80**, 245403 (2009)

6. S. Wippermann, N. Koch, S. Blankenburg, U. Gerstmann, S. Sanna, E. Rauls, A. Hermann, W. G. Schmidt, «*Understanding electron transport in atomic nanowires from large-scale numerical calculations*», High Performance Computing on Vector Systems '09, Springer-Verlag, S. 233-242

7. W. G. Schmidt, S. Blankenburg, E. Rauls, S. Wippermann, U. Gerstmann, S. Sanna, C. Thierfelder, N. Koch, M. Landmann, W. G. Schmidt, «*Understanding long-range indirect interactions between surface adsorbed molecules*», High Performance Computing in Science and Engineering '09, Springer-Verlag, S. 75-84

8. S. Chandola, K. Hinrichs, M. Gensch, N. Esser, S. Wippermann, W. G. Schmidt, F. Bechstedt, K. Fleischer, and J. F. Mc Gilp, «*Structure of Si(111)-In nanowires*

determined from midinfrared optical response», Phys. Rev. Lett. **102**, 226805 (2009), selected for Virtual Journal of Nanoscale Science & Technology 22(06) (2009)

9. S. Wippermann, W. G. Schmidt, «*Optical anisotropy of the In/Si(111)-(4×1)/-(8×2) nanowire array*», Surface Sci. **603**, 247–250 (2009)

10. S. Wippermann, W. G. Schmidt, «*Water adsorption on clean Ni(111) and p(2×2)-Ni(111)-O surfaces calculated from first principles*», Phys. Rev. B **78**, 235439 (2008)

11. S. Wippermann, N. Koch, W. G. Schmidt, «*Adatom-induced conductance modification of In nanowires: Potential-well scattering and structural effects*», Phys. Rev. Lett. **100**, 106802 (2008), selected for Virtual Journal of Nanoscale Science & Technology 17(12) (2008)

12. W. G. Schmidt, M. Albrecht, S. Wippermann, S. Blankenburg, E. Rauls, F. Fuchs, C. Rödl, J. Furthmüller, A. Hermann, «*LiNbO$_3$ ground- and excited-state properties from first-principles calculations*», Phys. Rev. B **77**, 035106 (2008)

13. S. Wippermann, W. G. Schmidt, A. A. Stekolnikov, K. Seino, F. Bechstedt, A. Calzolari, M. B. Nardelli, «*Quantum conductance of In nanowires on Si(111) from first principles calculations*», Surface Sci. **601**, 4045 (2007)

14. A. A. Stekolnikov, K. Seino, F. Bechstedt, S. Wippermann, W. G. Schmidt, A. Calzolari, M. B. Nardelli, «*Hexagon versus trimer formation in In nanowires on Si(111): Energetics and quantum conductance*», Phys. Rev. Lett. **98**, 026105 (2007), selected for Virtual Journal of Nanoscale Science & Technology 15(3) (2007)

15. W. G. Schmidt, S. Blankenburg, S. Wippermann, A. Hermann, P. H. Hahn, M. Preuss, K. Seino, F. Bechstedt, «*Anomalous water optical absorption: Large-scale first-principles simulations*», High Performance Computing in Science and Engineering '06, Springer-Verlag, S. 49–58

16. S. Blankenburg, S. Wippermann, T. Krüger, «*Ensemble Teleportation under suboptimal conditions*», Physica Scripta **74**, 190-196 (2006)

Acknowledgements

First and foremost I wish to thank my supervisor Prof. Wolf Gero Schmidt for his continuous support, countless discussions and his readiness to address even smaller problems at any time. His insights and sense of direction have been invaluable for this work.

I also wish to thank Prof. Arno Schindlmayr for very helpful discussions. Especially his insights into many-body perturbation theory are much appreciated.

Prof. Thomas Frauenheim's valuable advice and support is gratefully acknowledged. My work experience with him as a student sparked my interest in solid state physics.

Many thanks go to my coworkers Simone Sanna and Nadja Koch, who have been working with me on the nanowire project. I also wish to thank my current and former colleagues here at Paderborn University. Simone Lange, Uwe Gerstmann, Eva Rauls, Stephan Blankenburg, Christian Thierfelder, Andreas Hermann, Mark Landmann, Martin Rohrmüller, Björn Lange, Jan-Philip Hülshoff, Florian Schulte and Philipp Wette have always been a pleasure working with.

I am also very grateful to Prof. Friedhelm Bechstedt and his group at Jena University. Claudia Rödl, Frank Fuchs and Jürgen Furthmüller provided me with help and updated versions of their GW/BSE implementations many times.

I am indebted to Prof. Molinari, Prof. Nardelli, Arrigo Calzolari, Andrea Ferretti, and Rosa DiFelice as well for providing much appreciated support and advice in my transport calculations with the WanT code.

I also wish to thank Prof. Duan Wenhui and Prof. Jerry Bernholc for their hospitality and generosity when I visited them in Beijing and Raleigh, respectively. It has been a pleasure and was enlightening to stay and work in their groups. Many thanks go to Wang Peng, Huang Bing, Li Jia, Yan Binghai, Zou Xiao Long, Han Wei and Xu Yong in Beijing. A big thank you also goes to Lu Wen Chang for his help with the real-space code and Cecilia Upchurch in Raleigh.

Generous grants of computing time from the Paderborn Center for Parallel Computing (PC2) and the Höchstleistungsrechenzentrum Stuttgart (HLRS) are greatly appreciated. Many thanks go to Sabine Roller, Martin Bernreuther, Jens Simon, Axel Keller and all those other people there for providing technical assistance and making things work.

Finally I wish to thank my family for their invaluable support. I am especially indebted to my wife Han Fei for her endless patience, encouragement, keeping my back free of everything else and providing me with self-cooked food in the office so many times.

谢谢你的爱

I want morebooks!

Buy your books fast and straightforward online - at one of world's fastest growing online book stores! Environmentally sound due to Print-on-Demand technologies.

Buy your books online at
www.morebooks.shop

Kaufen Sie Ihre Bücher schnell und unkompliziert online – auf einer der am schnellsten wachsenden Buchhandelsplattformen weltweit! Dank Print-On-Demand umwelt- und ressourcenschonend produziert.

Bücher schneller online kaufen
www.morebooks.shop

KS OmniScriptum Publishing
Brivibas gatve 197
LV-1039 Riga, Latvia
Telefax: +371 686 204 55

info@omniscriptum.com
www.omniscriptum.com

Printed by Books on Demand GmbH, Norderstedt / Germany